IEE Materials & Devices Series 6

Series Editors: Dr N. Parkman
Professor D. V. Morgan

OPTICAL FIBRE

Other volumes in this series

1. Insulating liquids: their uses, manufacture and properties
   A. C. M. Wilson
2. Electrical insulation
   A. Bradwell (Editor)
3. Gallium arsenide for devices and integrated circuits
   H. Thomas, D. V. Morgan, B. Thomas, J. E. Aubrey and
   G. B. Morgan (Editors)
4. Semiconductor lasers for long-wavelength optical-fibre
   communications systems
   M. J. Adams, A. G. Steventon, W. J. Devlin and
   I. D. Henning
5. Semiconductor device modelling
   C. M. Snowden

OPTICAL FIBRE

by

C.K. Kao

Peter Peregrinus Ltd. on behalf of The Institution of Electrical Engineers

Published by: Peter Peregrinus Ltd., London, United Kingdom

British Library Cataloguing in Publication Data

Kao, Charles K. (Charles Kuen)
Optical fibre.
1. Fibre optics
I. Title II. Series
621.36'92

ISBN 0 86341 125 8

Printed in England by Short Run Press Ltd, Exeter

Contents

Preface

They say that fluency in any language is only achieved once the usage of prepositions has been mastered. Alas, this is something that still eludes my husband, the author of this book. I was, therefore, called upon to read and edit the drafts of his manuscript, and as my reward was invited to write this preface.

This book is based on a series of lectures that were given as a graduate course at Yale University, New Haven, USA, first in 1984, and then again in 1986 and in 1987. Using the experience gained from the first round of lectures, the second and third rounds were progressively modified, re-structured and more finely honed. My husband takes delight in teaching students, more so when they are responsive and lively. The Yale students themselves suggested some of the changes and they pointed out the areas which were difficult to grasp. I take this opportunity to thank them for their interest in this field, an area in which the author has spent the greater part of his career. We both wish them well in their future endeavours, wherever they may be. I also take this opportunity to thank the many people, who by allowing items from their publications to be reproduced, contributed to the clarity and usefulness of the text.

Readers of this book should have a relatively broad scientific background; at least to initial university science subjects levels. The material deals with the relevant physics and chemistry of fibre optic systems and confines itself entirely to the theoretical basis of the practical aspects of the fibre alone. This it does in a fairly exhaustive manner. Used as a text book, it would be ideal for senior year university students or for post-graduate courses. For engineers working in the field of fibre optics, this book is a practical monograph that would be useful as a reference and which would add to their understanding of the subject.

In 1985, for his pioneering contribution to this field of opto-electronics, my husband, Charles Kao, was elected a Marconi Fellow. At that time we were in Stuttgart, West Germany, where he was spending a year working at Standard Electrik Lorenz, a subsidiary then of ITT Corporation. The phone call came from the US at, for them, a perfectly respectful hour. The time for us was in the small hours of the morning. My husband answered the bedside phone, muttered a thank you very much, dropped the phone and promptly fell sound asleep again. Mrs. Braga, I can assure you, after I shook him to full wakefulness, we were too excited to sleep further that night! We remain very conscious of the honour you and your committee have bestowed upon him.

Much thought was given to ways of executing the commission required from every Marconi Fellow; a commission that gives renewed incentive to each Fellow to pursue creative work, aimed at enhancing the quality of human lives through improved means of communication. With such thoughts in mind, we hope this book will be useful to students world-wide. We dedicate this work to the spirit of Guglielmo Marconi.

Thanks are given to the Marconi Prize Committee for the esteem and high honour they have given to this eleventh Marconi Fellow and for the inspiration they continue to give to those working at the leading edge in telecommunications. In 1988, it will be our pleasure to attend the fourteenth Marconi Prize Ceremony in Rome, Italy, when the author will report on how his commission was accomplished.

Gwen M. W. Kao
Hong Kong
May 1988

Chapter 1

A history of the development of Optical Fibre

At every stage of development of the light-guiding optical fibre, some level of scientific understanding is involved. It is a fascinating example to use, to illustrate what learning is all about.

We start with free-space propagation of light. Before the discovery that light is a form of electromagnetic energy whose characteristics are described by Maxwell's equations, much empirical knowledge was gained through careful experimentation. It was discovered that light appeared to propagate in a straight line through air, and to be reflected by surfaces and diffracted by transparent solids or liquids. A geometrical theory of light was then developed which described the light propagation path through various media. According to this theory, light rays propagate in a straight line from A to B in free space. The law of reflection states that a ray that impinges on a surface is reflected by the surface such that the angle of incidence is equal to the angle of reflection. The angles are measured relative to the normal of the surface (Fig. 1.1).

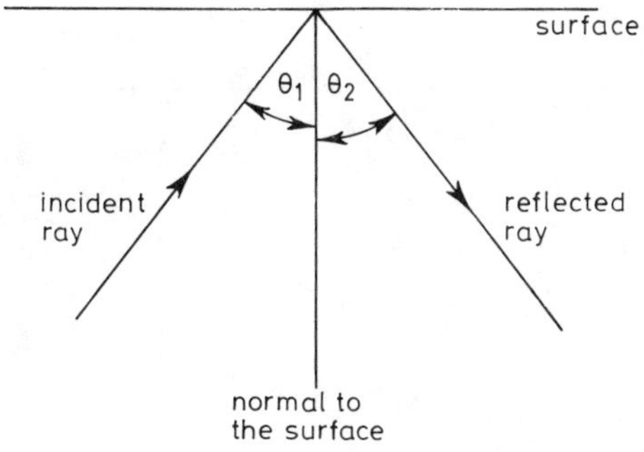

Fig. 1.1 *Law of reflection*

The law of refraction states that a ray undergoes refraction when it propagates from one medium across an interface to a second medium such that the ratio of the sine of the incident angle and the sine of the refracted angle is the inverse of the relative refractive index of the two media. The angles are measured relative to the normal of the interface (Fig. 1.2).

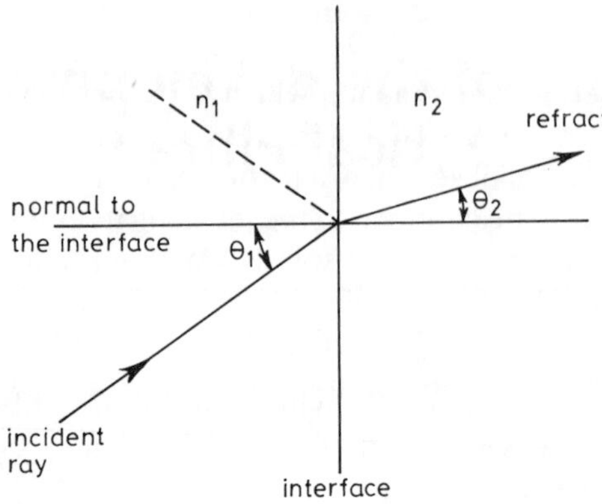

Fig. 1.2 *Law of refraction*

$$\frac{\sin\theta_1}{\sin\theta_2} = \frac{n_2}{n_1}$$

If $n_1 > n_2$ then at $\theta_1 = \theta_c$, the critical angle for total internal reflection, no refracted ray emerges, but total internal reflection takes place according to the law of reflection.

From this simple theory, it can be seen that a ray propagating from a dense medium through successive layers of rarer media will undergo bending and eventually total internal reflection. Hence, in a continuously varying medium, ray bending and total internal reflection is expected since such a medium can be approximated by successive layers of decreasing refractive indices (Fig. 1.3).

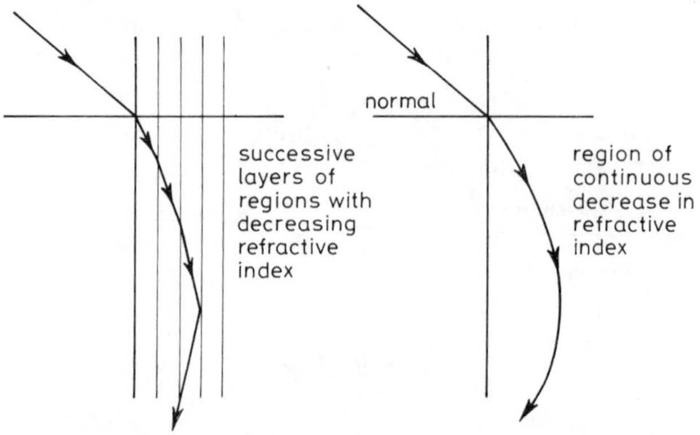

Fig. 1.3 *Ray propagation through successive layers of rarer media*

Total internal refraction can easily be observed in nature. One example is to see the shining silver-like surface of the swimming-pool surface when viewed at certain angles from a submerged position. The other example is to see the mirage effect while driving on the highway on a hot, humid day. In this case the air near the surface of the road is heated by the hot road to a higher temperature, and hence is at a lower density than the air further away from the surface. Thus the driver is observing from a denser medium towards a rarer medium.

It is the total reflection phenomenon which is responsible for trapping the light within the optical fibre. That light rays are guided by a column of transparent material can be observed in a jet of water illuminated at one end by light. The light appears to be confined to the jet along its entire length and only released at the other end with a splash. Without additional scientific knowledge, an optical fibre capable of guiding light can be envisaged. What is needed is a thread of transparent material of higher refractive index than air. Light can be launched in one end, and it should emerge from the other end.

Glass is an obvious candidate material for optical-fibre light guides. It is transparent, and it can be pulled into a thread of uniform cross-section. Indeed, glass fibres with diameters less than 1/4 mm are also highly flexible. A light guide capable of bending light appears possible. However, the steps towards developing a light guide with prescribed performance meeting practical application requirements are many and involve several scientific and technology breakthroughs.

1.1 Flexible light pipes

One of the earlier applications is the flexible light pipe. The aim is to allow visible light to be guided to inaccessible spots via a tortuous path, usually well under 1 m in length. Assuming for the moment that transparent material is available and uniform structures can be made, the first two issues are: (i) how to support the fibre without causing undue loss of light and (ii) what curvature can the light follow. The first issue can be serious.

It is clear from the basic laws of optics that the fibre can only guide light if the incidence angle is along the axial direction and up to angle $(90—\theta_c)°$ relative to the axial direction for meridian rays, i.e. rays which cross the fibre axis. The magnitude of the angle can be shown geometrically to be dependent on the critical angle, and hence, in turn dependent on the relative refractive indices of the fibre and its surrounding medium (Fig. 1.4).

$$\text{Since } \frac{\sin\theta}{\sin(90-\theta_c)} = \frac{n_1}{n_2} = n \text{ and } \sin\theta_c = \frac{1}{n}$$

$$\text{Therefore } \sin\theta = (n^2-1)^{\frac{1}{2}}$$

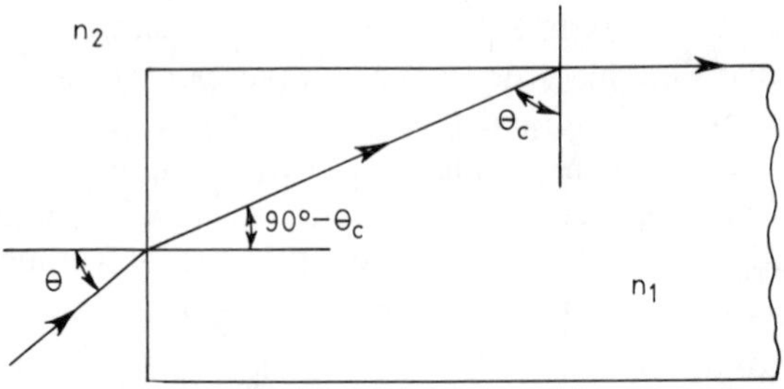

Fig. 1.4 *Basic theory of light pipes*

sin θ is called the numerical aperture (NA) of the fibre with $n = n_1/n_2$; hence if $n_2 = 1$ for air then $n = n_1$. If the fibre in air is touched by a substance with a higher refractive index than air, the numerical aperture of the fibre will decrease. In other words, some rays which would be guided by the fibre will now be refracted out of the fibre. In the limit when $n_2 = n_1$, all rays will not be guided. In addition, if the fibre is surrounded by an absorbing medium, the guided light will be progressively attenuated. Hence, a practical fibre must have a light guiding core enclosed in a coaxial cladding of a second transparent material whose index is lower than the core. The larger the index difference, the larger the angle of acceptance. For light pipes, the criteria for efficient and large light carrying capacity are:

- large core diameter
- high core refractive index
- low cladding refractive index
- sufficient cladding thickness
- good flexibility
- good transmission efficiency

These criteria are qualitative and served initially as engineering design yardsticks. The transparency of glass for optical components such as lenses and prisms was found to be around 10 dB per metre. This means that light is attenuated by a factor of 2 over a distance of about 1 ft which is found to be reasonably adequate. The cladding material thickness of several micrometres was found to be sufficient to isolate the fibre guiding surface. For flexibility, it was found that the total fibre diameter should be less than 0·3 mm. In fact, the challenge at the beginning was to find two glasses with sufficient index differences that have fairly close thermal-expansion coefficients and transparency. It was found that one method to make the fibre is to put a glass rod inside a tube of cladding glass and heat a region of several centimetres of the combination structure until molten. By pulling the composite rod and tube, the molten sector will neck into a tapered region to form a fibre of a much smaller diameter. At a constant rate of drawing, a steady state is reached when the rod and tube are slowly fed into the hot zone and a fibre is drawn out such that the volume of material entering and leaving the hot zone is equal. This fibre-forming process requires the surface of the rod and inner surface of the tube to be as smooth as possible, and since the glass undergoes melting and re-solidification, the expansion coefficient of the core glass must be higher than the cladding so that the resulting fibre is strong and not brittle (see Chapter 5 for explanation).

The success of making such fibre light pipes generated sufficient interest for commercial exploitation. Many fibres, assembled in a bundled form, act as a good flexible light pipe. If the fibres are assembled in ordered rows and columns, image transmission can be achieved between the two ends. In other words, a picture placed in contact with one end of a fibre bundle will appear at the other end. Both ends must be flat and polished.

A plausible explanation of how light is guided around curved path can readily be derived. Using the basic laws of optics, it can be seen that, in a curved guide, a light ray incident normally to the fibre end surface will reach the critical angle for total internal reflection for a curvature equal to the critical angle. This means that light will escape when a straight fibre undergoes bending (Fig. 1.5).

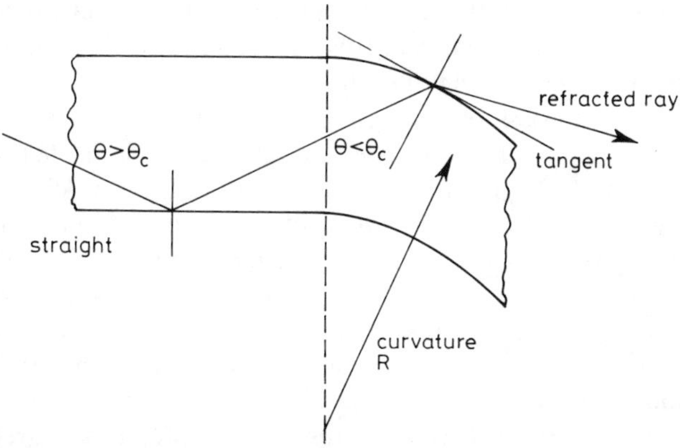

Fig. 1.5 *Guidance of light round a curved path*

In practice, for a fibre NA of 0·3 and a core diameter of 0·3 mm, with a radius of about 7 mm corresponding to a curvature angle of 16·7 °, light leakage will occur even if the fibre is initially carrying all rays parallel to the axis. It follows, for the rays propagating near critical angle, leakage will occur as soon as the fibre bends. Hence, when this type of fibre is bent to a radius of curvature of around 1 cm, strong radiation loss occurs.

1.2 Fibre as an electromagnetic waveguide

By the time light was identified as electromagnetic energy, Maxwell's equations governing the electromagnetic-wave behaviour were already well known. Plane wave in free space is a solution of the Maxwell's equations. Its physical characteristics can be verified in practice as electromagnetic energy in the form of radio waves or optical waves. The laws of optics can be derived formally in wave formalism. Polarisation of the optical plane wave is seen to be associated with the direction of the $\boldsymbol{E}$, electric field vector, or $\boldsymbol{H}$, the magnetic field vector. The $\boldsymbol{E}$ and $\boldsymbol{H}$ vectors in a plane wave are transverse and perpendicular to each other. Power is expressible in the form of a Poynting vector. The dielectric-waveguide problem was solved early in the 1900s, both for the planar infinite slab guide and for the circular rod surrounded by a second infinite medium. The solution is valid only for ideal dielectric material. It provides a means to determine the field distribution of propagating and radiative modes, as well as propagation constants. It shows that a finite number of modes can propagate in a dielectric structure corresponding to specific geometric ray directions. The radiative modes form a continuum and correspond to unguided refracted rays. The wave description provides an exact description of the characteristics of the fibre as a waveguide, and allows the fibre design to be improved.

The field distribution explains the role of the cladding clearly. For each mode the field distribution can be seen to be dependent on the operating wavelength and the dimensions of the waveguide, and the material refractive indices of the waveguide. The operating parameter is $V = (2\pi a/\lambda)(n_2^2 - n_1^2)^{\frac{1}{2}}$, which is a dimensionless parameter. For each mode, the operating V value determines how near or away from cut off the mode is. When a mode is operating far from cut off,the evanescent field at the core/cladding interface penetrates very little from the core, while near cut off, the evanescent field extends several wavelengths into the cladding. Thus, the required cladding thickness to prevent the waveguide propagation from being disturbed when the fibre is being handled can be determined.

The perturbation of the waveguide, such as non-uniformity in cross-sectional dimension, interface non-uniformity or abrupt dimensional changes such as those occurring when two fibres are joined together, are problems of significant theoretical difficulties. Much progress, however, has been made towards obtaining desired solutions. The dielectric loss of the waveguide is treated theoretically as a perturbation problem. The loss is assumed to be small so that it does not change the solution obtained for a structure of lossless dielectric material. With these solutions the characterisation of the fibre waveguide as far as propagation is concerned is fairly complete. The advantages and limitations of the light pipe are made reasonably clear.

The fibre is a dielectric waveguide suitable for guiding electromagnetic energy at optical wavelength. It can serve as a light pipe for visible light. A practical fibre can be made with two glasses, one with higher refractive index serving as the light-guiding core, and the other with lower refractive index, serving as the cladding which permits the fibre to be handled without disturbing the light propagation. It is highly flexible if the fibre overall diameter is less than 0·3 mm. Using lead glasses, an adequate amount of energy can be delivered over distances up to 1 m. This situation gave rise to two early applications for the fibres. One is as a light pipe for bringing light to regions through a non-straight path. Several thousands of fibres are bundled together to increase the total amount of light delivered. The second application is as an image-transmission device. If thousands of fibres are stacked so that the fibre ends are positioned in the same regular array at both ends, then an image presented at one end surface will appear at the other end surface. This arrangement is sometimes called fibre scope or image bundle. The image resolution depends on the size of the fibre core and the thickness of the cladding.

For these applications the problems to be addressed are mainly in the fibre making and handling. The propagation characteristics are satisfactory. Fibres are to be drawn and cut to length. Bundles are to be formed and ends prepared. Fibre breakage, speed of production, polishing, diameter-size uniformity, and fibre stacking are some of the manufacturing problems. The breakage is particularly irksome since every broken fibre will cause a speck on the image-transmission bundle. This problem is alleviated somewhat by coating the fibre with some resin during the fibre-drawing process. This step was found to decrease the fragility of the fibre.

This was the situation in the 1950s for the field of fibre optics. When the interest in optical communication started in the late 1950s and early 1960s, fibre offered some apparent advantages as an optical waveguide. However, more information and understanding were needed to assess the real potentials. Fibre for operation as a transmission line at optical wavelength must have low loss and adequate information-carrying capacity. It must be practically handleable and should be manufacturable in long length and at a reasonable cost.

Foremost on the list of needed information was knowledge concerning fibre-transmission loss. The attenuation must be much lower than 10 dB/m or 10 000 dB/km. A maximum tolerable loss of 20 dB/km must be attained before the fibre can be considered for transmission application. An investigation into loss mechanisms shows that dielectric material has, in general, four types of losses. Two are more basic and are associated with the molecular and atomic structures of the materials. These are due to electronic and molecular vibrations which cause absorptions. The other two are associated with external influences. One is a scattering loss due to micro-inhomogeneity of the material and the other is an absorption loss due to contaminants. It became clear that, in the infra-red wavelengths, molecular vibrational losses are serious for polymeric materials which have strong infra-red absorptions. Crystalline and glassy material can be transparent up to and beyond 10 μm wavelength. The electronic vibrational bands for these materials are in the ultra-violet end of the electromagnetic spectrum. Generally, crystalline and glassy material are transparent in the visible and near-ultra-violet spectral regions.

The scattering losses were found to be high in crystalline material since dislocations are frequent, but low in single-crystal materials. It is the glassy materials which look most promising. Scattering losses seem low, and molecular and electronic absorption bands for SiO_2 based glass are far removed from the visible and near infra-red region. Hence, a glass fibre appears to be the best bet, especially since the fibre-making process from glass has already been developed for low-melting-point glasses, and the process is somewhat understood. Unfortunately, the glass loss is unacceptably high and the cause of the loss not well understood. The question 'Can glass loss be reduced from 10 000 dB/km to at most 20 dB/km?' needed to be answered.

Early work in making colour glasses indicate that transition elements such as Cu, Fe, Mn, Co absorb light strongly at certain wavelengths. These absorption bands can be broad or sharp. This suggests that the absorption loss of glass could be greatly reduced if the unwanted transition elements are removed. The spectral study of absorption bands of impurity elements in glass has already shown that the intensity of absorption is proportional to the concentration. By extrapolation, it is possible to determine that impurity levels below 1 part in 10^6 is needed to reduce absorption losses to below 20 dB/km in the 0·8 to 1 μm spectral region. The prospect of reducing the impurity level to less than 1 part in 10^6 is reasonable since chemicals have routinely been purified to better than 1 in 10^9 purity or 1 part per billion of impurity.

Nonetheless, purification of glass is less straightforward because of the glass-making process, this usually involves high temperatures and an uncontrolled environment. Furthermore, techniques for determining impurity levels in glass are not well developed. So, at the beginning it was difficult to assess whether systematic reduction of impurities would indeed achieve low loss. The calibration of the impurity absorption coefficient can only be determined with an impurity level of around 1000 parts in 10^6 and above. There was no assurance that the linear decrease of absorption with impurity concentration would continue linearly. Attempts to verify this met with difficulties both in estimation of the purity level as well as in determining the loss.

The impurity level in glass is measured by using various mass-spectroscopic methods, trace-element detection techniques and wet chemical methods. Each presented significant practical difficulties. In mass spectroscopy, the sensitivity is governed by the instrument, as well as the uncertainty of how the element is split into its valance states. In wet chemical methods the impurity of the solvents and containers could jeopardise the measurement accuracy.

Even if a reasonably good estimation had been obtained of the impurity level, the loss measurement of the actual glass presented a great deal of difficulties. First, when glass loss is only 100 dB/km, the total loss of a 100 cm sample is 0·1 dB. This is small compared with the reflection loss of the surface reflections. To try to measure 20 dB/km loss with a 10 cm sample is like trying to measure the size of a grain of sand with a ruler calibrated only in inches.

Special measurement equipment must be developed, or very long samples must be made. Both these approaches were pursued.

Spectroscopic-loss-measurement equipment capable of measuring differential losses of 0·01% of two glass samples with 10 cm length difference was constructed. Measurement on synthetic silica with <1 part in 10^6 total impurity verified that loss of the order of 20 dB/km or below is indeed attainable. Measurement on long uncladded fibre made from the silica also indicated that the loss is low. In this case, the accuracy of the measurement is affected by the cleanliness and uniformity of the sample, and by the inconsistent loss at the fibre supports.

With the loss issue reasonably resolved, the next priority task is to devise means to make fibre from materials of sufficient purity. At the same time the information-capacity and waveguide-design issue were reconsidered. It is important to note that the inquiry into loss mechanisms in materials required the disciplines of material science and measurements. Not only, is knowledge of chemistry involved in general, but also involved is an in-depth understanding of material properties, measurement techniques, interaction of electromagnetic waves with materials, atomic and molecular properties of solids, crystallography and glass sciences. Also development of new understanding in many of these subjects is needed; in particular, methods for impurity analysis with greatly improved sensitivity, and spectroscopic-loss-measurement techniques. It is also important to note, with hindsight, that the scientific and technological background did turn out to be adequate for the successful development of fibre technology. If the infrastructure of scientific theory and measurement instrumentation was not in place or could not be developed in time, then the quest towards a practically useful fibre would have been doomed from the start.

Careful analysis of the electromagnetic-waveguide properties of the dielectric waveguide provided the answers to determine the configuration of the designed waveguide structure for communication applications. It was shown that a waveguide can support many modes. For a particular V value, i.e. a specific set of values of relative refractive index, operating wavelength and fibre-core diameter, a finite number of modes will propagate without loss (in a lossless dielectric waveguide), while the others are beyond cut off and are not guided by the structure. All modes will propagate at their respective velocity and with their specific field distribution.

All modes have a cut off except the lowest order HE_{11} hybrid mode. A single-mode-operating condition is possible when $V < 2{\cdot}4$. This is a particularly important and attractive situation since wave propagation will take place at a uniquely defined configuration and with a single propagation coefficient. The question arose whether the physical dimensions are practical and handleable. For an operating wavelength of 1 μm and NA = 0·07 and $n_1 = 1{\cdot}5$ a core diameter of about 5 μm is needed. At first sight this appeared to be rather small. However, considering that the core is to be enclosed in a cladding of sufficient thickness, the physical waveguide with an OD of 100 μm or so would be convenient for handling, while the core diameter is unlikely to be too small for alignment to light sources, detectors or to each other. The overwhelming advantage lies in the intrinsic bandwidth or information-carrying capacity of such a waveguide. It was later shown to be easily greater than 100 GHz km. This type of waveguide can also be shown to be able to negotiate bends with negligible loss. Theoretically the bend is a difficult electromagnetic-wave problem, particularly for circular fibres. However, inferences from geometric-optics analysis and planar-waveguide analysis provided assurances that bending loss should be low, and this was later proved to be the case in practical situations. More refined theoretical analysis based on optical-ray analysis and formal electromagnetic solutions are now available to confirm this finding.

The research towards low-loss material was a significant undertaking. As material systems were examined it became obvious that high-temperature glasses were the most attractive. Simple silicate glasses and pure silica are among the best candidates. These have electronic and molecular absorption bands well removed from the region where suitable light sources exist, namely from visible to 2 μm wavelength.

Two approaches were almost simultaneously adopted. The first is to examine means to make quartz (pure silica) into a cladded fibre, since very pure quartz can be made using a flame-hydrolysis process or a plasma-induced chemical vapour-deposition process from halides of silicon. The second is to make lower-temperature melting silica-based glasses such a soda–lime–silicate and borosilicate glasses with ultra-high-purity starting chemicals in a contamination-free environment. The former has the advantage that the process of making high-purity quartz was already proven, but quartz is a glass with an exceptionally low refractive index. Therefore, a higher-index material with equal low loss must be developed and clad with pure silica, or a means to make a glass with lower refractive index than quartz must be found. Besides, fibre drawing of quartz is difficult because of the very high temperature involved ~2000 °C. The latter has greater flexibility in the choice of glass types, but the removal of impurities and/or the exclusion of new sources of contaminations during the glass-making process led to totally new challenges.

The evolution towards a successful process was long and tortuous. In the multi-element glass development, the first stage was to purify glass constituent materials in oxide and carbonate form. Simultaneously, the methods to handle the pure chemical during the drying, mixing, transport and storage processes must be carefully developed. Early lessons showed how easily contaminants can be introduced through a non-clean environment, containers, stirrers and a whole host of handling procedures. Later, the purity of raw materials which can be achieved through wet chemical means showed a tendency to reach a limit of around several parts per billion. The contamination problem at melting is probably the most challenging one. Since the temperature is elevated, contaminents can come through the furnace insulation, the crucible and the stirrer. Platinum crucibles contaminate the glass residue with the transition elements from the crucible material as well as Pt particles. Silica crucible with RF heating and oxygen bubbling as a means to stir the glass were tried. Human handling of the chemicals was reduced to a minimum and in a sealed environment. Eventually glass loss below 20 dB/km was attained. In fact, the best glass loss at 0·8 μm was less than 5 dB/km when multi-element glass was finally abandoned in favour of the lower-loss high-silica glasses.

The approach to making high-silica fibre was based on the chemical vapour-deposition principle. This involves the vaporisation of volatile components of glass-making elements and reacting them with oxygen. The vaporisation process purifies the raw material since transition elements have higher vapour pressure than silicon. Starting with pure halides, the silica formed can attain better than 1 part per billion purity. Silica glass in particulate form settles on a cool surface. In a suitable mechanical arrangement this pure glass can be collected in such a way as to form the fibre core. In one arrangement the vapour is carried into a hot zone within a tube by a stream of oxygen gas. The reactants are deposited on the inner surface of the tube and gradually fused. Then the tube, which is usually made of pure quartz, is heated until it collapses into a rod, entrapping the deposited material to form the fibre core. The core material is a mixture of pure silica and germania. The latter increases the refractive index of the glass. Several variants of the process were developed for fibre production in commercial volumes. Further development is still taking place with a view to reducing production and material costs.

In the early stages of development, the kinetics of the reactions were not understood. For example, by bubbling oxygen through silicon tetrachloride and then heating the vapour, SiO_2 was formed. Similarly with oxygen and germanium tetrachloride, GeO_2 should be formed. However, if both are taking place, the resulting mixed Ge-Si-O glass will have different compositions depending on the reaction-zone temperature and length. In the case of fluorine incorporation, the process could yield glass with no fluorine incorporation unless the reaction condition is favourable.

After the composite rod is formed, fibre is drawn from this 'preform' in the fibre-drawing process. The high softening temperature of silica glass makes fibre drawing difficult. The furnace systems capable of reaching around 2000 °C tended to fall into two types—DC driven or RF driven. They had graphite or zirconia heating elements and were expensive and had short service life. Initially it was easier to use oxy-hydrogen flame torches as the heat source for fibre drawing.

After the fibre is formed (it is important also to keep it from contaminants as it is formed in the hot zone) its surface is perfect, but must be immediately protected with a plastic or resin coating. The application of fibre coating is a significant engineering task. The choice of material is not arbitrary. The coating material should have reasonable hardness, abrasion resistance, be strippable, contain no abrasive material, and it is sometimes desirable for it to have a higher refractive index than the cladding. The viscosity should be reasonable, so that a suitable applicator can be designed. The material should be heat or ultra-violet curable, and contain no solvent.

The surface perfection of the fibre and fibre strength are closely linked. The mechanical characteristics of fibres are subjected to intensive studies, as tough practical requirements for fibre strength and durability are needed for different applications. It opens up yet another scientific and technological field for investigation. Glass is a special solid with linear stress/strain relationship holding theoretically up to 20% elongation. The malleability is zero and it breaks as a brittle solid. The break can be traced to the rupture of the Si–O–Si bond. For glass with flaws in the form of cracks, the mechanical characteristics are very different. It is governed by the stress at the crack tip. If the environment is corrosive, i.e. it can react with glass, the crack will propagate even if the tip stress is well below the bond fracture stress. This leads the crack tip stress to increase, and hence an acceleration in the crack-propagation speed. When the tip stress reaches fracture stress, catastrophic failure takes place. This process is known as fatigue.

The study of the durability of fibre and how to improve durability leads to the study of how to make the fibre hermetic. The passivation of a fibre surface can be achieved by coating the fibre with hermetic material such as SiN_2 or metal. The chemistry and mechanics involved are still under study. Incidentally, hermeticity is a way to prevent hydrogen from diffusion into the fibre causing attenuation increase.

The fabrication of fibre calls for material-science considerations. For a material-related problem this is not unexpected. However, since the waveguide transmission characteristics are intimately associated with the optical properties of the material, the scientific consideration involves both material science and physics. It turns out that the dielectric constant or refractive index is frequency dependent. This is known as material dispersion. For silica the material dispersion is opposite in sign to the waveguide dispersion. In fact, it is fortuitous that the magnitudes of dispersion are equal and opposite at 1·3 μm for a germanium-silicate core fibre designed for practical single-mode operation with a simple step-index core. This means that, operating at 1·3 μm, the fibre waveguide has zero dispersion, and therefore has infinite information-carrying capacity.

The successful development of a single-mode fibre was preceded by a significant period during which the prospect of achieving the target loss and the availability of a suitable light source at a desirable wavelength and size were all in doubt. During this period, early fibre geometry involved a step-index core of about 50 μm and an OD of 125 μm. Owing to the large core size, the number of modes it can support are large. The dispersion due to the mode-velocity differences limits the bandwidth to several tens of MHz km. The light source available was LEDs at 0·9 μm with spectral width of about 600 Å. The waveguide loss was about 10 dB/km. Hence, a link length of 2 km can be attained. This was the situation when the first operating fibre system was installed.

The next development was a fibre with a graded-index core. Based on a geometric-optics argument, the modal dispersion is due to the path-length difference between the rays travelling along the fibre axis and those zigzag rays. Using the same rationale, the modal dispersion can be compensated by grading the index such that the shortest path has the highest index, and therefore the propagation velocity will be slower than the zigzag rays propagating in lower-index regions. It can be shown that a near-parabolic index distribution according to $n = n_0(1 - ar^2)$ can equalise the velocity of all axial modes. At the same time, the fabrication technique can readily be adapted to produce such graded-index fibres. This type of fibre is now standardised with nominal NA of 0·2 core diameter of 50 μm and OD of 125 μm.

Analysis based on electromagnetic theory shows how the profile of index distribution can be optimised for zero dispersion at a given operating wavelength. The bandwith–distance product for such a waveguide can be consistently larger than 1000 MHz km. The fabrication technology by then was sufficiently well developed that the fibre loss was not contributed to by any of the impurity absorption except the OH^- ion, which has a strong absorption band with its second-harmonic absorption bands at 1·39 μm. Otherwise, the waveguide loss is due to scattering loss of the bulk material only. Since scattering loss is inversely proportional to wavelength to the fourth power, the loss at longer wavelength is significantly less than that at 0·85 μm. The best loss of graded-index fibre is about 3 dB/km at 0·85 μm, and about 1 dB/km at 1.3 μm. Automated fabrication facilities also reduced fibre cost by improving yield. The availability of lasers operating both at 0·85 μm and 1·3 μm gives rise to the possibility of using the fibre to carry two signals simultaneously on each optical carrier The fibre is therefore required to have adequate bandwidth at both these operating wavelengths. Such a fibre is sometimes called a two-window fibre, and fibre specifications often call for a bandwidth–distance product of, say, 800 MHz km at both 0·85 μm and 1·3 μm. This type of fibre is not easy to make, since the profile must be accurately controlled so that the minimum dispersion occurs in between the two operating wavelengths. At the same time the relatively large core and high NA of this type of fibre calls for a substantial amount of GeO_2, which is a high-cost material. This prompted some work on alternative dopants to be investigated. Aluminium (Al), was considered. This involves using an Al compound which is in a solid form. The fibre-making process is made somewhat more complicated.

The whole issue of dopant-material cost vanished when single-mode fibre returned as the favourite fibre design for general telecommunications applications. This occurred as a result of the confidence gained in dealing with the smaller core size and the excellent loss characteristics and bandwidth available. Since single-mode fibres operate with a NA of less than 0·1, and the core diameter is 10 times smaller than that of the multimode fibre, the amount of dopant material needed is much reduced. The reduced NA also meant lower scattering loss, so that, at 1·3 μm, loss is as low as 0·4 dB/km and, at 1·55 μm, it is less than 0·2 dB/km. Repeater spacing can readily be greater than 50 km, making repeaterless operation possible in most areas.

The dispersion characteristics of single-mode fibre is excellent when operated with a light source having narrow spectral width. At 1·3 μm the zero dispersion of the fibre permits the spectal width of the light source to be relatively broad. At other operating wavelengths the waveguide dispersion or material dispersion gives rise to a dispersion of around 10 ps/nm km, limiting the repeater distance for a high-bit-rate signal by its dispersion rather than the loss. If single-frequency light sources are used, adequate bandwidth permits transmission-bandwidth–distance products up to 1000 GHz km. If the single-mode fibre core is profiled or segmented, the zero-dispersion wavelength can be shifted. These dispersion-shifted fibres can be made with small loss increase. Trade-off exists between improving the laser source to maintain single-frequency operation, or improving the fibre, to provide zero-dispersion performance, to match the source spectral width. The choice is application dependent and is far from being clear cut.

Single-mode fibre has well defined characteristics. Unlike multimode system, the operational characteristics are highly stable. However, a single-mode fibre with a circular cross-section can be considered as a two mode structure, since two orthogonal polarisations are permitted. If the core is not exactly circular, the two modes will have different velocities. This results in reduced bandwidth and certain operational difficulties owing to the power interchange between the two modes. For coherent wave propagation, a polarisation-maintaining or single-polarisation type of single-mode fibre may be needed. The former can be achieved by keeping the core geometry constant and also by minimising stress-induced birefringence. The latter can be obtained by deliberately making the fibre elliptical or through stressed-induced birefringence.

The necessity to join, connect and couple fibre stimulates progress in techniques to establish mechanical precision of the order of a fraction of a micrometre. The kinematic design and the choice of suitable material so that submicron alignment precision can be maintained over a wide temperature are challenging scientific and technological problems, especially if the method has to be cost effective as well.

With fibre being almost of zero loss, near infinite bandwidth and close to zero cost, the application range is broadening beyond the transmission-line fibre. It is creating opportunities as a means of creating new physical sensors and signal-processing elements. For these applications a fibre is likely to be operating with coherent light sources and detection techniques. The stage has arrived when fibre-optical waveguide must be considered in the same way that microwave waveguides have been treated.

Fibre components for optical-wave manipulation similar to those available for millimetric and microwave can be envisaged. This requires a thorough understanding of the fibre taper and coupled fibres. Furthermore, the choice of materials for fibre components is much wider, since relatively higher losses of the order of 10—100 dB/km are perfectly acceptable. Moreover, with appropriate dopant and optical pumping energy, active fibre components can also be envisaged. Indeed, a fibre laser at 1·06 μm was demonstrated a while ago.

Material nonlinearity is also an important factor for single-mode fibre, particularly when operating with coherent optical waves. The desirability for fibre to carry higher power is evident for increasing repeater spacing in transmission applications and as a means to transmit optical power. For high-power operation the power density within the fibre can be high enough to cause material breakdown. For most glasses this can occur when power density reaches about 10^8 W/cm^2. For a single-mode waveguide carrying 10 mW power the power density already reaches 10^6 W/cm^2. This is a safe power level, but some nonlinear effects can be induced. Moreover, within a single-mode waveguide, the nonlinear effects can be cumulative, so that even with 1 mW of coherent power induced, nonlinear scattering can be significant.

A general understanding of optical nonlinear effects in glass also helps to define whether nonlinearity can serve a useful function, as well as setting operational conditions for the avoidance of detrimental effects. Examples of the usefulness of nonlinearity can be found in optical bistability which may serve as very high-speed optical logic or as switching elements.

The non-linearity-induced refractive-index change can result in modification of propagation characteristics of fibre waveguide. This can be used on one hand to produce power-sensitive couplers and, on the other hand, to attain conditions suitable for soliton propagation. Solitons propagate with zero dispersion, which hence implies perhaps infinite information capacity over very long distances.

In the course of developments, optical fibres summoned the efforts of people from a broad range of scientific and technological disciplines, often requiring them to develop interdisciplinary skill and interest. Electromagnetic theory and material science are basic and must be applied broadly and in very considerable depth; opening up many new challenging extensions of existing state of the art. New measurement instruments and techniques must be developed. New understanding of underlying theoretical issues must be established. This applies particularly as the push towards improving fabrication techniques and strength and durability intensifies. The radiation hardness, non-linearity, new glassy materials and EM solutions for waveguides with varying cross-section and profiled index distribution all can result in new products whose full benefits can only be assessed through a clear identification of the technological and scientific limits.

Future requirements and possibilities are promoting efforts in searching for material systems suitable for fibre applications with ultimate theoretical losses of 0·001 dB/km. This has a chance of being achieved for operation in the 2–10 μm spectal region. Non-linearity phenomena and their relationship with material structure is promoting an interest in broadly directed studies towards non-linear crystalline, glassy and semiconductor materials. Integrated optoelectronic components and operational concepts also promise to have profound implications in future communications/information-processing systems. These, however, extend the fibre-optics field to an area where the presence or absence of fibre is almost incidental. Nevertheless, integration of ideas is essential as an insurance against sub-optimal realisation of products.

The presence of optical fibre as the ideal transmission medium is a precursor to the development of a new improved communication network needed for our information society. The implication of this is far reaching. It is reminiscent of how a motorway system transformed inland transportation and how large high-speed jet aircraft changed worldwide transportation. A communication revolution is in the making.

Chapter 2

Electromagnetic theory of fibre waveguides

The wave propagation of electromagnetic energy can be described by Maxwell's equations. Specific solutions have been obtained for many important practical problems in waveguide and antenna areas. However, problems with analytic solutions are only obtained when the geometry and boundary conditions are favourable. The dielectric-waveguide or fibre problem had been solved for some idealised cases, namely, the planar and cylindrical waveguide cases early in the 20th century. The classical approach starts with the set of Maxwell's equations for a linear, isotropic, lossless dielectric material having no current sources and free charges:

$$\nabla \times \boldsymbol{E} = -\frac{\partial \boldsymbol{B}}{\partial t} \tag{2.1}$$

$$\nabla \times \boldsymbol{H} = \frac{\partial \boldsymbol{D}}{\partial t} \tag{2.2}$$

$$\nabla . \boldsymbol{D} = 0 \tag{2.3}$$

$$\nabla . \boldsymbol{B} = 0 \tag{2.4}$$

with

$$\boldsymbol{D} = \varepsilon\varepsilon_0 \boldsymbol{E} \tag{2.5}$$

$$\boldsymbol{B} = \mu\mu_0 \boldsymbol{H} \tag{2.6}$$

Using the vector

$$\nabla \times (\nabla \times \boldsymbol{A}) = \nabla(\nabla . \boldsymbol{A}) - \nabla^2 \boldsymbol{A} \tag{2.7}$$

and eqs. 2.3—2.6 and taking $\nabla \times$ of eqns. 2.1 and 2.2,

$$\nabla^2 \boldsymbol{E} = \varepsilon\varepsilon_0 \mu\mu_0 \frac{\partial^2 \boldsymbol{E}}{\partial t^2} \tag{2.8}$$

$$\nabla^2 \boldsymbol{H} = \varepsilon\varepsilon_0 \mu\mu_0 \frac{\partial^2 \boldsymbol{H}}{\partial t^2} \tag{2.9}$$

which are known as wave equations. Note μ and ε are assumed to be constants.

$$\nabla^2 \boldsymbol{E} = \frac{\varepsilon\mu}{c^2} \frac{\partial^2 \boldsymbol{E}}{\partial t^2} \tag{2.10}$$

$c = 1/\sqrt{\varepsilon_0\mu_0}$ is the speed of light in free space.

For a cylindrical waveguide the wave equation in cylindrical co-ordinate is

$$\frac{1}{r}\frac{\partial}{\partial r}\left(r\frac{\partial \boldsymbol{E}}{\partial r}\right) + \frac{1}{r^2}\frac{\partial^2 \boldsymbol{E}}{\partial \phi^2} + \frac{\partial^2 E}{\partial^2 z^2} = \frac{\mu\varepsilon}{c^2}\frac{\partial^2 \boldsymbol{E}}{\partial t^2} \tag{2.11}$$

with a similar equation for $\boldsymbol{H}$.

We seek a solution for a wave with frequency f, $\omega = 2\pi f$, and propagating in the z direction with propagation constant β.

The solution can proceed in a number of different ways, one of which is to show by using eqns. 2.1 and 2.2 that the transverse components of the $\boldsymbol{E}$ and $\boldsymbol{H}$ fields can be expressed in terms of the axial components E_z and H_z, which satisfies the wave equations

$$\frac{\partial^2 E_z}{\partial r^2} + \frac{1}{r}\frac{\partial E_z}{\partial r} + \frac{1}{r^2}\frac{\partial E_z}{\partial \phi^2} + (\omega^2\varepsilon\mu - \beta^2)E_z = 0 \tag{2.12}$$

Applying the separable variable method for solving partial differential equations, due to circular symmetry, $\boldsymbol{E}_z$ is of the form:

$$E_z \propto F_1(r) e^{j\nu\phi} e^{j(\omega t - \beta z)} \tag{2.13}$$

The equation reduces to

$$\frac{\partial^2 F_1}{\partial r^2} + \frac{1}{r}\frac{\partial F_1}{\partial r} + \left(\omega^2\varepsilon - \beta^2 - \frac{\nu^2}{r^2}\right)F_1 = 0 \tag{2.14}$$

Designating $u^2 = \omega^2\varepsilon - \beta^2$ and recognizing this to be the Bessel equation

$$E_z = AJ_\nu(ur) e^{j\nu\phi} e^{j(\omega t - \beta z)} \tag{2.15}$$

Applying the boundary conditions that the tangential $\boldsymbol{E}$ and $\boldsymbol{H}$ components at the waveguides boundary are continuous and the field must vanish at infinity, an eigenfunction can be established. The solution of this equation for β represents the possible modes. The equation is complicated and is only amenable to numerical solutions. The propagating modes can be classified into those with $\boldsymbol{E}$ field transverse to the direction of propagation and those with both $\boldsymbol{E}$ and $\boldsymbol{H}$ field components in the direction of propagation. The former is known as TE modes while the latter are designated HE and EH modes. The cut-off conditions of all modes can be derived without explicitly solving the eigenfunction.

For our discussion, we follow the analysis rationalised by A. W. Snyder and his co-workers.[1] This approach is particularised for optical-fibre waveguides and caters for a spatial varying n in the x,y plane.

We assume that the fibre has a geometry, refractive indices and co-ordinate system as shown in Fig. 2.1, where n_{co} is the maximum value of core refractive index, and n_{cl} is the value of the uniform cladding index.

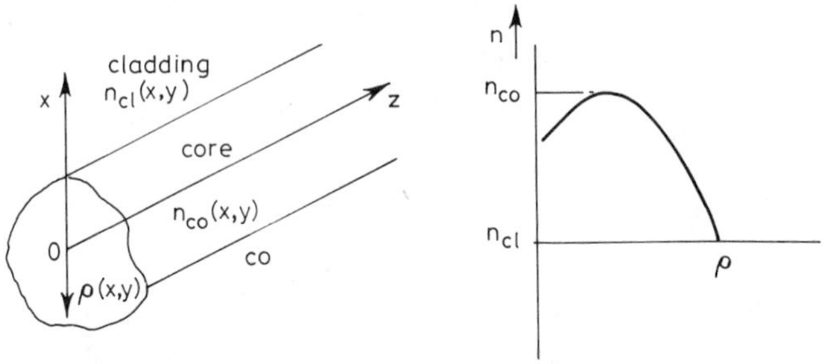

Fig. 2.1 *Optical-fibre waveguide with spatial varying n in x, y plane*

The $\boldsymbol{E}$ and $\boldsymbol{H}$ fields are each to be expanded in two parts, the discrete guided and the continuous radiative mode, as follows:

$$\boldsymbol{E}(x,y,z) = \sum_j a_j\boldsymbol{E}_j(x,y,z) + \sum_j a_{-j}\boldsymbol{E}_{-j}(x,y,z) + \boldsymbol{E}_{rad}(x,y,z) \qquad (2.16)$$

$$\boldsymbol{H}(x,y,z) = \sum_j a_j\boldsymbol{H}_j(x,y,z) + \sum_j a_{-j}\boldsymbol{H}_{-j}(x,y,z) + \boldsymbol{H}_{rad}(x,y,z) \qquad (2.17)$$

The guided modes are source-free solutions to Maxwell's equations for the waveguide.

This formulation allows single and multimode waveguides to be analysed and some discontinuity problems to be tackled.

The properties of guided modes and their field are now derived from the Maxwell equations. In order to establish that the transverse field can be expressed in terms of the longitudinal component, we set

$$\boldsymbol{E} = (\boldsymbol{E}_t + e_z\hat{z}) \exp \mathrm{f01}(\beta z - \omega t) \qquad (2.18)$$

$$\boldsymbol{H} = (\boldsymbol{H}_t + h_z\hat{z}) \exp j(\beta z - \omega t) \qquad (2.19)$$

with $\boldsymbol{E}_t$ and $\boldsymbol{H}_t$ the transverse components and e_z and h_z the longitudinal components. The solutions for the longitudinal components are sought. The solutions are valid for waveguides with no z variations when the field is expressible as a superposition of $\boldsymbol{E}$ and $\boldsymbol{H}$. and z is the propagating direction. The propagation constant and the wave are periodic in time with angular frequency ω. Hence, $\omega = 2\pi f$. We take $\varepsilon_0\mu_0$ as free-space dielectric constants ε, μ as relative constants for permitivity $\varepsilon = \varepsilon_r\varepsilon_0$ or $n^2\varepsilon_0$ and permeability $\mu = \mu_r\mu_0$ where n is the refractive index. These are functions of the transverse co-ordinates. We have $1/\sqrt{\varepsilon_0\mu_0} = c$, the velocity of light in free space. We take $k = 2\pi/\lambda$ to be the free-space wave number, where λ is the wavelength; thus $f\lambda = c$ in free space and $k = 2\pi f/c = \omega/c$

Omitting writing from now on, for the time dependence, which will remain implicit, and for the sourceless case:

$$\nabla \times \boldsymbol{E} = (\mu_0/\varepsilon_0)^{\frac{1}{2}}\, k\, \boldsymbol{H} \tag{2.20a}$$

$$\nabla \times \boldsymbol{H} = -\mathrm{i}(\varepsilon_0/\mu_0)^{\frac{1}{2}}\, kn^2\boldsymbol{E} \tag{2.20b}$$

$$\nabla\cdot(n^2\boldsymbol{E}) = 0 \tag{2.20c}$$

$$\nabla\cdot\boldsymbol{H} = 0 \tag{2.20d}$$

Substituting eqns. 2.18 and 2.19 into eqn. 2.20 (*a* and *d*) and using the expansion

$$\nabla \times \boldsymbol{A} = \nabla_t \times \boldsymbol{A}_t + i\beta\hat{z} \times \boldsymbol{A}_t - \hat{z} \times \nabla_t A_z$$

For example, eqn. 2.20*b*

$$\nabla \times \boldsymbol{H} = -\, i(\varepsilon_0/\mu_0)^{\frac{1}{2}}\, kn^2\boldsymbol{E}$$

becomes

$$\nabla_t \times \boldsymbol{H}_t + i\beta\hat{z} \times \boldsymbol{H}_t - \hat{z} \times \nabla_t h_z = -\, i(\mathrm{t}_0/\mu_0)^{\frac{1}{2}} kn^2(\boldsymbol{E}_t + \hat{z}e_z)$$

Equating transverse components

$$\boldsymbol{E}_t = (\mu_0/\varepsilon_0)^{\frac{1}{2}}\, \frac{1}{kn^2}\, \hat{z} \times (\beta\boldsymbol{H}_t + i\nabla_t h_z) \tag{2.22}$$

and equating longitudinal component while taking $\hat{z}$. product.

$$\boldsymbol{e}_z = i\,(\mu_0/\varepsilon_0)^{\frac{1}{2}}\frac{1}{kn^2}\,\hat{z}\cdot\nabla_t \times \boldsymbol{H}_t \tag{2.23}$$

From eqn. 2.20*c*

$$\nabla\cdot(n^2\boldsymbol{E}) = 0 \quad \text{If } n = n(x, y)$$

using $\nabla\cdot\boldsymbol{A} = \nabla_t\cdot\boldsymbol{A}_t + i\beta A_z$ (2.24)

$$\nabla_t(n^2\boldsymbol{E}_t) + i\beta n^2 e_z = 0$$

Therefore

$$e_z = \frac{i}{\beta}\frac{1}{n^2}\left\{ \nabla t\cdot(n^2\boldsymbol{E}_t) \right\}$$

$$= \frac{i}{\beta}\frac{1}{n^2}\left\{ n^2\nabla_t\cdot\boldsymbol{E}_t + \boldsymbol{E}_t\cdot\nabla_t n^2 \right\}$$

$$= \frac{i}{\beta}\left\{ \nabla_t\cdot\boldsymbol{E}_t + (\boldsymbol{E}_t\cdot\nabla_t)\ln n^2 \right\} \tag{2.25}$$

Note

$$\frac{\partial}{\partial x}\ln n^2 = \frac{\partial}{\partial n}\ln n^2 \frac{\partial n}{\partial x}$$

$$= \frac{1}{n^2}\,2n\,\frac{\partial n}{\partial x}$$

$$= \frac{1}{n^2}\frac{\partial}{\partial x}(n^2)$$

Hence

$$\nabla \ln n^2 = \frac{1}{n^2}\nabla n^2 \tag{2.26}$$

Similarly, we have

$$\boldsymbol{H}_t = \left(\frac{\varepsilon_0}{\mu_0}\right)^{\frac{1}{2}}\frac{1}{k}\hat{z}\times\left\{ \beta\boldsymbol{E}_t + i\nabla_t e_z \right\} \tag{2.27}$$

$$h_z = i\left(\frac{\varepsilon_0}{\mu_0}\right)^{\frac{1}{2}}\frac{1}{k}\hat{z}\cdot\nabla_t\times\boldsymbol{E}_t = \frac{i}{\beta}\nabla_t\cdot\boldsymbol{H}_t \tag{2.28}$$

Eliminating $\boldsymbol{H}_t$ and $\boldsymbol{E}_t$ in turn from eqns. 2.22 and 2.27 by substitution, $\boldsymbol{E}_t$ and $\boldsymbol{H}_t$ can be seen to be expressible in terms of the tangential components only:

$$\boldsymbol{E}_t = \frac{1}{k^2n^2-\beta^2}\left\{ \beta\nabla_t e_z - (\mu_0/\varepsilon_0)^{\frac{1}{2}}k\hat{z}\times\nabla_t h_z \right\} \tag{2.29}$$

$$\boldsymbol{H}_t = \frac{i}{k^2n^2-\beta^2}\left\{ \beta\nabla_t h_z + (\varepsilon_0/\mu_0)^{\frac{1}{2}}kn^2\hat{z}\times\nabla_t e_z \right\} \tag{2.30}$$

$(k^2n^2 - \beta^2)$ is the wave number (2.31)

These expressions are valid for $n = n(x,y)$ and allow the transverse components to be derived once the longitudinal solutions are found from the wave equations.

The wave equation without source terms is called a homogeneous wave equation. For fields with separable transverse and longitudinal form as discussed here, the vector wave equations are:

$$(\nabla_t^2 + n^2k^2 - \beta^2)\,\boldsymbol{E} = -\,(\nabla_t + i\beta\hat{z})\,\boldsymbol{E}_t\cdot\nabla_t \ln n^2 \tag{2.32}$$

$$(\nabla_t^2 + n^2k^2 - \beta^2)\boldsymbol{H} = \{(\nabla_t + i\beta\hat{z}) \times \boldsymbol{H}_t\} \times \nabla_t \ln n^2 \tag{2.33}$$

2.1 Step-profiled waveguides

This class of waveguides has a uniform core surrounded by a finite uniform cladding which is assumed to be unbounded. The core refractive index is n_{co} and that of the cladding is n_{cl}. For certain geometries exact analytical solutions can be constructed.

Since $\nabla_t \ln(n^2)$ is now zero in the core and cladding region, the equation to be satisfied is

$$(\nabla_t^2 + n^2k^2 - \beta^2)e_z = 0 \tag{2.34}$$

$$(\nabla_t^2 + n^2k^2 - \beta^2)h_z = 0 \tag{2.35}$$

For planar waveguide with co-ordinates as shown in Fig. 2.2

$$n(x) = n_{co} \text{ for } 0 \leqslant |x| \leqslant \rho$$

$$n(x) = n_{cl} \text{ for } \rho < |x| < \infty$$

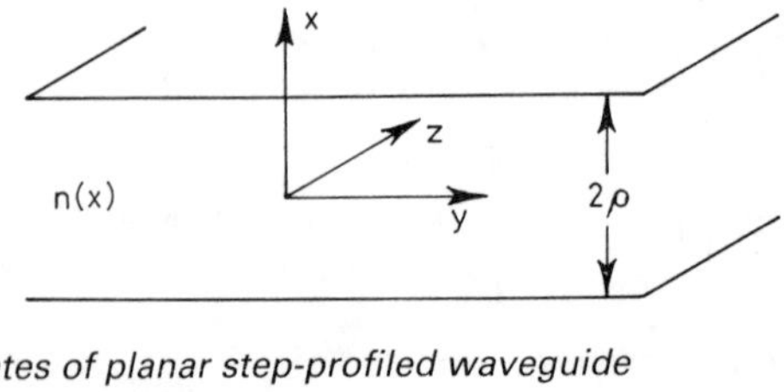

Fig. 2.2 *Co-ordinates of planar step-profiled waveguide*

By expanding eqns. 2.29 and 2.30 in Cartesian co-ordinates

$$e_x = \frac{i}{p}\left\{\beta\frac{\partial e_z}{\partial x} + \left(\frac{\mu_0}{\varepsilon_0}\right)^{\frac{1}{2}} k\frac{\partial h_z}{\partial y}\right\} \tag{2.36a}$$

$$e_y = \frac{i}{p}\left\{\beta\frac{\partial e_z}{\partial y} - \left(\frac{\mu_0}{\varepsilon_0}\right)^{\frac{1}{2}} k\frac{\partial h_z}{\partial x}\right\} \tag{2.36b}$$

$$h_x = \frac{i}{p}\left\{\beta\frac{\partial h_z}{\partial x} - \left(\frac{\varepsilon_0}{\varepsilon_0}\right)^{\frac{1}{2}} kn^2\frac{\partial e_z}{\partial y}\right\} \tag{2.36c}$$

$$h_y = \frac{i}{p}\left\{\beta\frac{\partial h_z}{\partial y} + \left(\frac{\varepsilon_0}{\mu_0}\right)^{\frac{1}{2}} kn^2\frac{\partial e_z}{\partial x}\right\} \tag{2.36d}$$

$$\text{where } p = k^2n^2 - \beta^2 \tag{2.37}$$

Hence all transverse components can be derived from the individual longitudinal components.

For a planar waveguide the e_z and h_z equations are decoupled; hence e_z or h_z can be assumed to be zero arbitrarily, leading to TM or TE modes, respectively. Since $\boldsymbol{E}$ and $\boldsymbol{H}$ have no y dependence:

TE case

Take $e_z = 0$ then

$$e_x = 0 \tag{2.38a}$$

$$e_y = \frac{i}{p}\left\{\left(\frac{\mu_0}{\varepsilon_0}\right)^{\frac{1}{2}} k\frac{\partial h_z}{\partial x}\right\} \tag{2.38b}$$

$$h_x = \frac{i}{p}\left\{\beta\frac{\partial h_z}{\partial x}\right\} \tag{2.38c}$$

$$h_y = 0 \tag{2.38d}$$

TM case

Take $h_z = 0$ then

$$e_x = \frac{i}{p}\left\{\beta\frac{\partial e_z}{\partial x}\right\} \tag{2.39a}$$

$$e_y = 0 \tag{2.39b}$$

$$h_x = 0 \tag{2.39c}$$

$$h_y = \frac{i}{p}\left\{\left(\frac{\varepsilon_0}{\mu_0}\right)^{\frac{1}{2}} kn^2\frac{\partial e_z}{\partial x}\right\} \tag{2.39d}$$

and we have

$$\boldsymbol{E} = \boldsymbol{E}_t \exp(i\beta z) \tag{2.40a}$$

$$\boldsymbol{H} = \boldsymbol{H}_t \exp(i\beta z) \tag{2.40b}$$

Let $h_z = 0$ and substituting $\boldsymbol{E}$ into eqn. 2.34 for the core region

$$\left(\frac{\partial^2}{\partial x^2} + n_{co}^2 k^2 - \beta^2\right) e_z = 0$$

Define $U = \rho\,(n_{co}^2 k^2 - \beta^2)^{\frac{1}{2}}$ (2.41)

$$\left(\rho^2 \frac{d^2}{dx^2} + U^2\right) e_z = 0 \tag{2.42}$$

Taking into consideration the boundary condition e_z can be a sine or cosine function.

Take the case

$$e_z = A \cos(Ux/\rho) \tag{2.43}$$

For the cladding $n = n_{cl}$, taking into consideration again the boundary condition

$$\left(\frac{d^2}{dx^2} + n_{cl}^2 k^2 - \beta^2\right) e_z = 0 \tag{2.44}$$

Defining $W = \rho\,(\beta^2 - n_{cl}^2 k^2)^{\frac{1}{2}}$ (2.45)

$$\left(\rho^2 \frac{d^2}{dx^2} - W^2\right) e_z = 0 \tag{2.46}$$

Let e_z be of the form $\exp(-W\,|\,x\,|/\rho)$ (2.47)

The expression for e_x becomes

$$e_x = -\frac{i\rho\beta A}{U} \sin\frac{Ux}{\rho} \quad \text{in the core} \tag{2.48}$$

To facilitate later computational needs, set $e_x = 1$ at $x = \rho$

Therefore

$$A = \frac{-U}{i\rho\beta \sin u} \tag{2.49}$$

Hence

$$e_x = \frac{\sin \dfrac{Ux}{\rho}}{\sin u} \quad \text{in core} \tag{2.50}$$

In the cladding

$$h_y = -\frac{i}{p}\left(\frac{\varepsilon_0}{\mu_0}\right)^{\frac{1}{2}} k\, n_{cl}^2 B \frac{W}{\rho}\, \frac{x}{|x|} \exp\left(-W|x|/\rho\right) \tag{2.51}$$

at $x = \rho$, h_y in cladding $= h_y$ in core, matching h_y at core/cladding boundary. Therefore

$$\frac{1}{\beta}\left(\frac{\varepsilon_0}{\mu_0}\right)^{\frac{1}{2}} k\, n_{co}^2 = -\frac{i}{p}\left(\frac{\varepsilon_0}{\mu_0}\right)^{\frac{1}{2}} k\, n_{cl}^2 \frac{W}{\rho} e^{-W} B$$

$$B = \frac{1}{\beta}\,\frac{n_{co}^2}{n_{cl}^2}\,\frac{p}{-i}\,\frac{\rho}{We^{-W}} \tag{2.52}$$

$$h_y = -\frac{i}{p}\left(\frac{\varepsilon_0}{\mu_0}\right)^{\frac{1}{2}} k\, n_{cl}^2 \cdot \frac{1}{\beta}\,\frac{n_{co}^2}{n_{cl}^2}\,\frac{p}{-i}\,\frac{\rho}{We^{-W}}\,\frac{W}{\rho}\,\frac{x}{|x|}\, e^{\frac{W|x|}{\rho}}$$

Therefore

$$h_y = \frac{1}{\beta}\left(\frac{\varepsilon_0}{\mu_0}\right)^{\frac{1}{2}} k\, n_{co}^2 \frac{\exp(W|x|/\rho)}{\exp(-W)}\left(\frac{x}{|x|}\right) \text{ in cladding} \tag{2.53}$$

To derive the eigenfunction, we match tangential fields at the core/cladding interface for e_x and h_y without setting A and B to specific values; we have

$$e_z = A\cos\left(Ux/\rho\right) \text{ in core}$$
$$= B\exp\left(-W|x|/\rho\right) \text{ in cladding}$$

$$h_y = -\frac{i}{p}\left(\frac{\varepsilon_0}{\mu_0}\right)^{\frac{1}{2}} k\, n_{co}^2 \frac{U}{\rho} \sin\left(\frac{Ux}{\rho}\right) A \quad \text{in core}$$

$$= -\frac{i}{p}\left(\frac{\varepsilon_0}{\mu_0}\right)^{\frac{1}{2}} k\, n_{cl}^2 \frac{W}{\rho} \exp\left(-W|x|/\rho\right) B \text{ in cladding}$$

$$p^2 = \rho^2\left(k^2 n^2 - \beta^2\right) = U^2/\rho^2 \text{ or } W^2/\rho^2$$

At $x = \rho$ we have two simultaneous equations for A and B. For a solution the determinant of the coefficients must equal zero.

$$\begin{vmatrix} \frac{-i[\ \ ]}{U} n_{co}^2 \sin U & \frac{-i[\ \ ]}{W} n_{cl}^2 e^{-W} \\ \cos U & e^{-W} \end{vmatrix} = 0$$

Hence $n_{co}^2\, W = -\, n_{cl}^2\, u\cot U$ (2.54)

It is found to be convenient to define a dimensionless parameter V.

$$V = k\rho\left(n_{co}^2 - n_{cl}^2\right)^{\frac{1}{2}} = k\rho n_{co}\left(2\Delta\right)^{\frac{1}{2}} = k\rho n_{co}\sin\theta_c \tag{2.55}$$

with $\Delta = \dfrac{n_{co}^2 - n_{cl}^2}{2n_{co}^2} = \dfrac{\sin^2\theta_c}{2}$

since $U = \rho\left(k^2 n_{co}^2 - \beta^2\right)^{\frac{1}{2}}$

$W = \rho\left(\beta^2 - k^2 n_{cl}^2\right)^{\frac{1}{2}}$

$$U^2 + W^2 = \rho^2 k^2\left(n_{co}^2 - n_{cl}^2\right) = V^2 \tag{2.56}$$

Table 2.1 *Modal field components for step-profile planar waveguide* (Reproduced from Ref. 1)

	e_y		h_x		h_z	
	Core	Cladding	Core	Cladding	Core	Cladding
Even TE	$\frac{\cos(UX)}{\cos U}$	$\frac{\exp(-W\|X\|)}{\exp(-W)}$	$-\frac{\beta}{k}\left(\frac{\varepsilon_0}{\mu_0}\right)^{\frac{1}{2}}\frac{\cos(UX)}{\cos U}$	$-\frac{\beta}{k}\left(\frac{\varepsilon_0}{\mu_0}\right)^{\frac{1}{2}}\frac{\exp(-W\|X\|)}{\exp(-W)}$	$\frac{iW}{k\rho}\left(\frac{\varepsilon_0}{\mu_0}\right)^{\frac{1}{2}}\frac{\sin(UX)}{\sin U}$	$\frac{iW}{k\rho}\left(\frac{\varepsilon_0}{\mu_0}\right)^{\frac{1}{2}}\frac{X}{\|X\|}\frac{\exp(-W\|X\|)}{\exp(-W)}$
Odd TE	$\frac{\sin(UX)}{\sin U}$	$\frac{X}{\|X\|}\frac{\exp(-W\|X\|)}{\exp(-W)}$	$-\frac{\beta}{k}\left(\frac{\varepsilon_0}{\mu_0}\right)^{\frac{1}{2}}\frac{\sin(UX)}{\sin U}$	$-\frac{\gamma}{k}\left(\frac{\varepsilon_0}{\mu_0}\right)^{\frac{1}{2}}\frac{X}{\|X\|}\frac{\exp(-W\|X\|)}{\exp(-W)}$	$\frac{iW}{k\rho}\left(\frac{\varepsilon_0}{\mu_0}\right)^{\frac{1}{2}}\frac{\cos(UX)}{\cos U}$	$\frac{iW}{k\rho}\left(\frac{\varepsilon_0}{\mu_0}\right)^{\frac{1}{2}}\frac{\exp(-W\|X\|)}{\exp(-W)}$

$e_x = e_z = h_y = 0$

	e_x		h_y		e_z	
	Core	Cladding	Core	Cladding	Core	Cladding
Even TM	$\frac{\cos(UX)}{\cos U}$	$\frac{n_{co}^2}{n_{cl}^2}\frac{\exp(-W\|X\|)}{\exp(-W)}$	$\frac{kn_{co}^2}{\beta}\left(\frac{\varepsilon_0}{\mu_0}\right)^{\frac{1}{2}}\frac{\cos(UX)}{\cos U}$	$\frac{kn_{co}^2}{\beta}\left(\frac{\varepsilon_0}{\mu_0}\right)^{\frac{1}{2}}\frac{\exp(-W\|X\|)}{\exp(-W)}$	$-\frac{IW}{\rho\beta}\frac{n_{co}^2}{n_{cl}^2}\frac{\sin(UX}{\sin U}$	$-\frac{iW}{\rho\beta}\frac{n_{co}^2}{n_{cl}^2}\frac{X}{\|X\|}\frac{\exp(-W\|X\|)}{\exp(-W)}$
Odd TM	$\frac{\sin(UX)}{\sin U}$	$\frac{n_{co}^2}{n_{cl}^2}\frac{X}{\|X\|}\frac{\exp(-W\|X\|)}{\exp(-W)}$	$\frac{kn_{co}^2}{\beta}\left(\frac{\varepsilon_0}{\mu_0}\right)^{\frac{1}{2}}\frac{\sin(UX)}{\sin U}$	$\frac{kn_{co}^2}{\beta}\left(\frac{\varepsilon_0}{\mu_0}\right)^{\frac{1}{2}}\frac{X}{\|X\|}\frac{\exp(-W\|X\|)}{\exp(-W)}$	$-\frac{iW}{\rho\beta}\frac{n_{co}^2}{n_{cl}^2}\frac{\cos(UX)}{\cos U}$	$-\frac{iW}{\rho\beta}\frac{n_{co}^2}{n_{cl}^2}\frac{\exp(-W\|X\|)}{\exp(-W)}$

$e_y = h_x = h_y = 0$

Note: $X = 1$ at $x = \rho$

Table 2.2: *Modal properties of the step-profile planar waveguide* (Reproduced from Ref. 1)

	Even TE modes	Odd TE modes	Even TM modes	Odd TM Modes
Eigen value equation	$W = U \tan U$	$W = -U \cot U$	$n_{co}^2 W = n_{cl}^2 U$ and U	$n_{co}^2 W = -n_{cl}^2 U \cot U$
Normalization	$N = \frac{\rho\beta}{2k}\left(\frac{\varepsilon_0}{\mu_0}\right)^{\frac{1}{2}} \frac{V^2}{U^2} \frac{1+W}{W}$		$N = \frac{\rho k n_{co}^2}{2\beta}\left(\frac{\varepsilon_0}{\mu_0}\right)^{\frac{1}{2}} \left\{1 + \frac{n_{co}^4}{n_{cl}^2}\frac{W^2}{U^2} + \frac{n_{co}^2}{n_{cl}^2}\frac{V^2}{U^2 W}\right\}$	
Fraction of power in core	$\eta = 1 - \frac{U^2}{V^2(1+W)}$		$\eta = 1 - \frac{n_{co}^2 n_{cl}^2 U^2}{n_{co}^2 n_{cl}^2 V^2 + n_{co}^4 W^3 + n_{cl}^4 W U^2}$	
Group velocity	$v_g = \frac{c}{n_{co}^2} \frac{\beta}{k} \frac{1}{1 - 2\Delta(1-\eta)}$			
Number of bound modes	$M_b = \mathrm{Int}\left(\frac{4V}{\pi}\right)$			

	TE_j *modes*	TM_j *modes*
Cutoff $U = V, W = 0$	$U = V = j\frac{\pi}{2}$	

All field components for the modes and the modal properties can now be expressed in terms of *U, W and V*. They are listed in Tables 2.1 and 2.2 Numerical solutions for the case with $n_{co} = 2{\cdot}5$, $n_{cl} = 1{\cdot}5$ are shown in Fig. 2.3

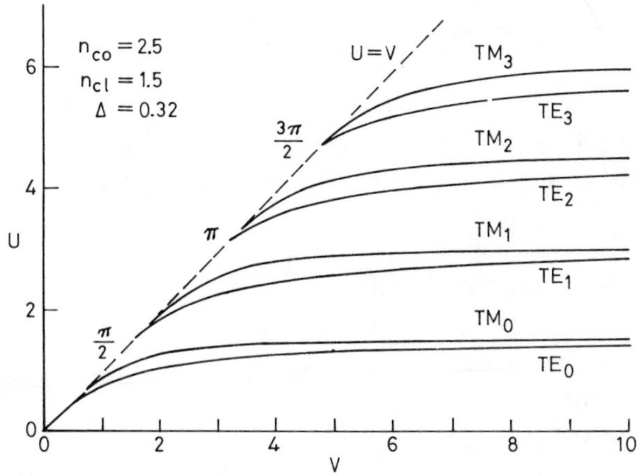

Fig. 2.3 *Numerical solution of the eigen value equations of Table 2.2 for the first eight modes*
The values along the dashed line are the cutoff values of *U*. (Reproduced from Ref. 1)

2.2 Step index fibre

The case of a step-index-core fibre with geometry as shown in Fig. 2.4

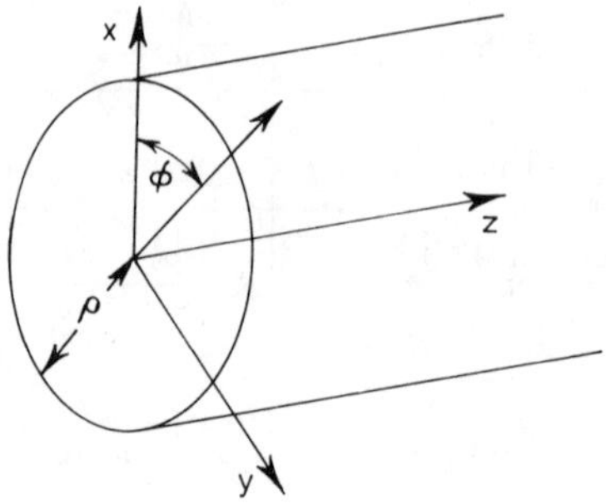

Fig. 2.4 *Geometry of Step-bidex-core fibre*

$$n(r) = n_{co} \quad 0 \leq r < \rho$$

$$n(r) = n_{cl} \quad \rho < r < \infty$$

We seek a solution with

$$\boldsymbol{E}(r,\phi,z) = \boldsymbol{E}_t(r,\phi) \exp{(i\beta z)}$$

$$\boldsymbol{H}(r,\phi,z) = \boldsymbol{H}_t(r,\phi) \exp{(i\beta z)}$$

The procedure is the same as in the planar-waveguide case. We have the simplified wave equations in cylindrical co-ordinate system.

$$\left\{\frac{\delta^2}{\delta R^2} + \frac{1}{R}\frac{\delta}{\delta R} + \frac{1}{R^2}\frac{\delta^2}{\delta\phi^2} + U^2\right\} \psi = 0 \qquad (2.57)$$

for $0 \leq R < 1$

$$\left\{\frac{\delta^2}{\delta R^2} + \frac{1}{R}\frac{\delta}{\delta R} + \frac{1}{R^2}\frac{\delta^2}{\delta\phi^2} - W^2\right\} \psi = 0 \qquad (2.58)$$

for $1 < R < \infty$

with $R = r/\rho$ and ψ denotes e_z or h_z

These are Bessel equations, and using the boundary conditions the solutions are of the form

$$\frac{A\,J_\nu(UR)}{J_\nu(U)}\cos\nu\phi \quad \text{for } 0 \leqslant R < 1 \tag{2.59}$$

$$\frac{A\,K_\nu(WR)}{K_\nu(W)}\cos\nu\phi \quad \text{for } 1 < R < \infty \tag{2.60}$$

Where J_ν and K_ν are Bessel functions of the first and second modified kind.

$$J_m(z) = \frac{i^m}{2\pi}\int_0^{2\pi} \exp(-iz\cos\theta)\cos(m\theta)\,d\theta$$

$$K_\nu(z) = \int_0^{\infty} \exp(-z\cosh t)\cosh(\nu t)dt$$

A is an arbitrary constant which is chosen to be consistent with the weak-guidance approximation results.

The tangential components can be derived from eqns. 2.29 and 2.30 in cylindrical co-ordinate systems. By equating tangential fields at $r = \rho$ or $R = 1$ for e_ϕ and h_ϕ, the eigenfunction is

$$\left[\frac{J_\nu'(U)}{UJ_\nu(U)} + \frac{K_\nu'(W)}{WK_\nu(W)}\right]\left[\frac{J_\nu'(U)}{UJ_\nu(U)} + \frac{n_{cl}^2}{n_{co}^2}\frac{k_\nu'(W)}{WK\nu(W)}\right] = \left(\frac{\nu\beta}{kn_{co}}\right)^2\left(\frac{V}{UW}\right)^4 \tag{2.61}$$

for $HE_{\upsilon m}$ and $EH_{\upsilon m}$ modes

This is a complicated equation even for numerical solutions. In order to gain a theoretical understanding, a simpler approach must be found. Fortunately for optical fibres, the refractive indices n_{co} and n_{cl} differ only by a small amount so that Δ is small, typically 0·01. This leads to much simplified approximate solutions. The field components of various modes and their modal properties are tabulated in Tables 2.3—2.6.

Table 2.3 *Field components of (a) $HE_{\nu m}$ and $EH_{\nu m}$ and (b) TE_{om} and TM_{om} modes of step-profile-fibre* (Reproduced from Ref. 1)

Component	Core	*Cladding*
e_r	$-\dfrac{a_1 J_{\nu-1}(UR) + a_2 J_{\nu+1}(UR)}{J_\nu(U)} f_\nu(\phi)$	$-\dfrac{U}{W}\dfrac{a_1 K_{\nu-1}(WR) - a_2 K_{\nu+1}(WR)}{K_\nu(W)} f_\nu(\phi)$
e_ϕ	$-\dfrac{a_1 J_{\nu-1}(UR) + a_2 J_{\nu+1}(UR)}{J_\nu(U)} g_\nu(\phi)$	$-\dfrac{U}{W}\dfrac{a_1 K_{\nu-1}(WR) - a_2 K_{\nu+1}(WR)}{K_\nu(W)} g_\nu(\phi)$
e_z	$\dfrac{-iU}{\rho\beta}\dfrac{J_\nu(UR)}{J_\nu(U)} f_\nu(\phi)$	$\dfrac{-iU}{\rho\beta}\dfrac{K_\nu(WR)}{K_\nu(W)} f_\nu(\phi)$
h_r	$\left(\dfrac{\varepsilon_0}{\mu_0}\right)^{\frac{1}{2}} \dfrac{kn_{co}^2}{\beta}\dfrac{a_3 J_{\nu-1}(UR) - a_4 J_{\nu+1}(UR)}{J_\nu(U)} g_\nu(\phi)$	$\left(\dfrac{\varepsilon_0}{\mu_0}\right)^{\frac{1}{2}} \dfrac{kn_{co}^2}{\beta}\dfrac{U}{W}\dfrac{a_5 K_{\nu-1}(WR) + a_6 K_{\nu+1}(WR)}{K_\nu(W)} g_\nu(\phi)$
h_ϕ	$-\left(\dfrac{\varepsilon_0}{\mu_0}\right)^{\frac{1}{2}} \dfrac{kn_{co}^2}{\beta}\dfrac{a_3 J_{\nu-1}(UR) - a_4 J_{\nu+1}(UR)}{J_\nu(U)} f_\nu(\phi)$	$-\left(\dfrac{\varepsilon_0}{\mu_0}\right)^{\frac{1}{2}} \dfrac{kn_{co}^2}{\beta}\dfrac{U}{W}\dfrac{a_5 K_{\nu-1}(WR) - a_6 K_{\nu+1}(WR)}{K_\nu(W)} f_\nu(\phi)$
h_z	$-i\left(\dfrac{\varepsilon_0}{\mu_0}\right)^{\frac{1}{2}} \dfrac{UF_2}{k\rho}\dfrac{J_\nu(UR)}{J_\nu(U)} g_\nu(\phi)$	$-i\left(\dfrac{\varepsilon_0}{\mu_0}\right)^{\frac{1}{2}} \dfrac{UF_2}{k\rho}\dfrac{K_\nu(WR)}{K_\nu(W)} g_\nu(\phi)$

$f_\nu(\phi) = \begin{cases} \cos(\nu\phi) \\ \sin(\nu\phi) \end{cases}; \quad g_\nu(\phi) = \begin{cases} -\sin(\nu\phi) & \text{even modes} \\ \cos(\nu\phi) & \text{odd modes} \end{cases}$	$F_1 = \left(\frac{UW}{V}\right)^2 \frac{b_1 + (1-2\triangle)b_2}{\nu}; \quad F_2 = \left(\frac{V}{UW}\right)^2 \frac{\nu}{b_1 + b_2}$
$a_1 = \frac{(F_2 - 1)}{2}; \quad a_3 = \frac{(F_1 - 1)}{2}; \quad a_5 = \frac{(F_1 - 1 + 2\triangle)}{2}$ $a_2 = \frac{(F_2 + 1)}{2}; \quad a_4 = \frac{(F_1 + 1)}{2}; \quad a_6 = \frac{(F_1 + 1 - 2\triangle)}{2}$	$b_1 = \frac{1}{2U}\left\{\frac{J_{\nu-1}(U)}{J_\nu(U)} - \frac{J_{\nu+1}(U)}{J_\nu(U)}\right\}$ $b_2 = -\frac{1}{2W}\left\{\frac{K_{\nu-1}(W)}{K_\nu(W)} + \frac{K_{\nu+1}(W)}{K_\nu(W)}\right\}$

TE modes

Component	Core	Cladding
e_ϕ	$\frac{-J_1(UR)}{J_1(U)}$	$\frac{-K_1(WR)}{K_1(W)}$
h_r	$\left(\frac{\varepsilon_0}{\mu_0}\right)^{\frac{1}{2}} \frac{\beta}{k} \frac{J_1(UR)}{J_1(U)}$	$\left(\frac{\varepsilon_0}{\mu_0}\right)^{\frac{1}{2}} \frac{\beta}{k} \frac{K_1(WR)}{K_1(W)}$
h_z	$i\left(\frac{\varepsilon_0}{\mu_0}\right)^{\frac{1}{2}} \frac{U}{k\rho} \frac{J_0(UR)}{J_1(U)}$	$-i\left(\frac{\varepsilon_0}{\mu_0}\right)^{\frac{1}{2}} \frac{W}{k\rho} \frac{K_0(WR)}{K_1(W)}$
$e_r = e_z = h_\phi = 0$		

TM modes

Component	Core	Cladding
e_r	$\frac{J_1(UR)}{J_1(U)}$	$\frac{n_{co}^2}{n_{cl}^2} \frac{K_1(WR)}{K_1(W)}$
e_z	$\frac{iU}{\rho\beta} \frac{J_0(UR)}{J_1(U)}$	$\frac{-in_{co}^2}{n_{cl}^2} \frac{W}{\rho\beta} \frac{K_0(WR)}{K_1(W)}$
h_ϕ	$\left(\frac{\varepsilon_0}{\mu_0}\right)^{\frac{1}{2}} \frac{kn_{co}^2}{\beta} \frac{J_1(UR)}{J_1(U)}$	$\left(\frac{\varepsilon_0}{\mu_0}\right)^{\frac{1}{2}} \frac{kn_{co}^2}{\beta} \frac{K_1(WR)}{K_1(W)}$
$e_\phi = h_r = h_z = 0$		

Table 2.4 *Eigenvalue equations for the step-profile fibre.* (Reproduced from Ref. 1)

$HE_{\nu m}$ and $EH_{\nu m}$ modes	$\left\{\frac{J'_\nu(U)}{UJ_\nu(U)}+\frac{K'_\nu(W)}{WK_\nu(W)}\right\}\left\{\frac{J'_\nu(U)}{UJ_\nu(U)}+\frac{n_{cl}^2}{n_{co}^2}\frac{K'_\nu(W)}{WK_\nu(W)}\right\}=\left(\frac{\nu\beta}{kn_{co}}\right)^2\left(\frac{V}{UW}\right)^4$
Alternative form	$k^2n_{co}^2F_1=\beta^2F_2$
TE_{Om} modes	$\frac{J_1(U)}{UJ_0(U)}+\frac{K_1(W)}{WK_0(W)}=0$
TM_{Om} modes	$\frac{n_{co}^2J_1(U)}{UJ_0(U)}+\frac{n_{cl}^2K_1(W)}{WK_0(W)}=0$

Numerical solutions are shown in Figs. 2.5 and 2.6.

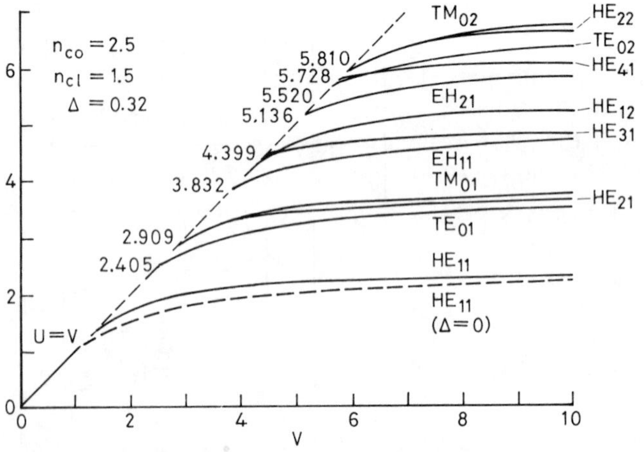

Fig. 2.5 *Numerical solution of the eigenvalue equations of Table 2.2 for the first 12 modes.*
The values along the dashed line are the cut-off values of U, and the dashed curve is the fundamental mode solution for a fibre with $n_{co}=n_{cl}$ and V fixed (Reproduced from Ref. 1)

2.3 Bounded modes

The propagation constant of the bounded mode is given by the β value for each mode or eigenvalue of the eigenfunction. Let β_j denote the propagation constant for the jth mode. The associated phase velocity is given by

$$v_j=\omega/\beta_j \tag{2.62}$$

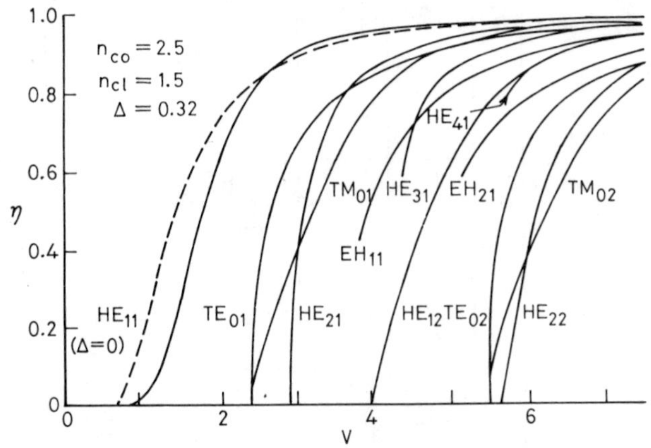

Fig. 2.6 *Fraction of modal power residing in the core for each of the first 12 modes as a function of V.*
The dashed curve is for the fundamental modes when$_{co}$ = n_{cl} and V fixed (Reproduced from Ref. 1)

for an operating frequency ω. Forward propagation is denoted by j with backward propagation by $-j$. All modes are orthogonal to each other and are orthogonal to the radiative field.

The power flow is described in terms of intensity S_j for jth modes, and is given by the magnitude of the time-averaged Poynting vector:

$$S_j = \tfrac{1}{2}|a_j|^2 \mathrm{Re}\{\boldsymbol{E}_j \times \boldsymbol{H}_j^* \cdot \hat{z}\} = \tfrac{1}{2}|a_j|^2 \boldsymbol{e}_j \times \boldsymbol{h}_j^* \cdot \hat{z} \tag{2.63}$$

Where

$$\boldsymbol{E}_j = \boldsymbol{e}_j \exp j\,(\beta z - \omega t)$$
$$\boldsymbol{H}_j = \boldsymbol{h}_j \exp j\,(\beta z - \omega t)$$

The total power is found by integrating S_j over the infinite cross-section A_∞:

$$P_j = \tfrac{1}{2}|a_j|^2 \int_{A_\infty} \boldsymbol{e}_j \times \boldsymbol{h}_j^* \cdot \hat{z}\, dA \tag{2.64}$$

Fraction of power in core is the ratio

$$n = \frac{\int_{A_{co}} \boldsymbol{e}_j \times \boldsymbol{h}_j^* \cdot \hat{z}\, dA}{\int_{A_\infty} \boldsymbol{e}_j \times \boldsymbol{h}_j^* \cdot \hat{z}\, dA} \tag{2.65}$$

The group velocity of the jth mode is the velocity of signal along the waveguide carried by the mode.

$$v_{gj} = \frac{d\omega}{d\beta_j} = -\frac{2\pi c}{\lambda^2}\frac{d\lambda}{d\beta_j} \tag{2.66}$$

$1/v_{gj} = \tau_j$ is referred to as group delay time. The transit time of signal carried by the jth mode over a distance z is

$$t = \frac{z}{v_{gj}} = z\tau_j = z\frac{d\beta_j}{d\omega} = -z\frac{\lambda^2}{2\pi c}\frac{d\beta_j}{d\lambda} \tag{2.67}$$

Table 2.5 *Modal properties of the step-profile fibre.* (Reproduced from Ref. 1)

		Core	Cladding
S_z	TE_{Om}	$\frac{\lvert a\rvert^2}{2}\left(\frac{\varepsilon_0}{\mu_0}\right)^{\frac{1}{2}}\frac{\beta}{k}\frac{J_1^2(UR)}{J_1^2(U)}$	$\frac{\lvert a\rvert^2}{2}\left(\frac{\varepsilon_0}{\mu_0}\right)^{\frac{1}{2}}\frac{\beta}{k}\frac{K_1^2(WR)}{K_1^2(W)}$
	TM_{Om}	$\frac{\lvert a\rvert^2}{2}\left(\frac{\varepsilon_0}{\mu_0}\right)^{\frac{1}{2}}\frac{kn_{co}^2}{\beta}\frac{J_1^2(UR)}{J_1^2(U)}$	$\frac{\lvert a\rvert^2}{2}\left(\frac{\varepsilon_0}{\mu_0}\right)^{\frac{1}{2}}\frac{kn_{co}^2}{\beta(1-2\Delta)}\frac{K_1^2(WR)}{K_1^2(W)}$
	$EH_{\nu m}$ *and* $HE_{\nu m}$	$\frac{\lvert a\rvert^2}{2}\left(\frac{\varepsilon_0}{\mu_0}\right)^{\frac{1}{2}}\frac{kn_{co}^2}{\beta J_\nu^2(U)}\left\{a_1a_3J_{\nu-1}^2(UR)\right.$ $+\,a_2a_4J_{\nu+1}^2(UR)$ $\left.\pm\frac{1-F_1F_2}{2}J_{\nu-1}(UR)J_{\nu+1}(UR)\cos(2\nu\phi)\right\}$	$\frac{\lvert a\rvert^2}{2}\left(\frac{\varepsilon_0}{\mu_0}\right)^{\frac{1}{2}}\frac{kn_{co}^2}{\beta K_\nu^2(W)}\frac{U^2}{W^2}\left\{a_1a_5K_{\nu-1}^2(WR)\right.$ $+\,a_2a_6K_{\nu+1}^2(WR)$ $\left.\pm\frac{1-2\Delta-F_1F_2}{2}K_{\nu-1}(WR)K_{\nu+1}(WR)\cos(2\nu\phi)\right\}$

$N = N_{co} + N_{cl}$	TE_{Om}	N_{co}	$\frac{\pi\rho}{2}\left(\frac{\varepsilon_0}{\mu_0}\right)^{\frac{1}{2}}\frac{\beta}{k}\left\{1 - \frac{J_0(U)J_2(U)}{J_1^2(U)}\right\}$	N_{cl}	$\frac{-\pi\rho^2}{2}\left(\frac{\varepsilon_0}{\mu_0}\right)^{\frac{1}{2}}\frac{\beta}{k}\left\{1 - \frac{K_0(W)K_2(W)}{K_1^2(W)}\right\}$
	TM_{Om}		$\frac{\pi\rho}{2}\left(\frac{\varepsilon_0}{\mu_0}\right)^{\frac{1}{2}}\frac{kn_{co}^2}{\beta}\left\{1 - \frac{J_0(U)J_2(U)}{J_1^2(U)}\right\}$		$\frac{-\pi\rho^2}{2}\left(\frac{\varepsilon_0}{\mu_0}\right)^{\frac{1}{2}}\frac{kn_{co}^2}{\beta(1-2\triangle)}\left\{1 - \frac{K_0(W)K_2(W)}{K_1^2(W)}\right\}$
	$\mathrm{EH}_{\nu m}$ and $\mathrm{HE}_{\nu m}$		$\frac{\pi\rho^2}{2}\left(\frac{\varepsilon_0}{\mu_0}\right)^{\frac{1}{2}}\frac{kn_{co}^2}{\beta J_\nu^2(U)}[a_1a_3\{J_{\nu-1}^2(U) - J_\nu(U)J_{\nu-2}(U)\} + a_2a_4\{J_{\nu+1}^2(U) - J_\nu(U)J_{\nu+2}(U)\}]$		$\frac{-\pi\rho^2}{2}\left(\frac{\varepsilon_0}{\mu_0}\right)^{\frac{1}{2}}\frac{kn_{co}^2}{\beta K_\nu^2(W)}\frac{U^2}{W^2}[a_1a_5\{K_{\nu-1}^2(W) - K_\nu(W)K_{\nu-2}(W)\} + a_2a_6\{K_{\nu+1}^2(W) - K_\nu(W)K_{\nu+2}(W)\}]$
η	All	$\frac{N_{co}}{N_{co} - N_{cl}}$			
v_g	All	$\frac{c}{n_{co}^2}\frac{\beta}{k}\frac{1}{1-2\triangle(1-\eta)}$			
M_{bm}		$\mathrm{Int}\left\{\frac{V^2}{2}\right\}: V \gg 1$			

S_z, N, η, v_g and M_{bm} are parameters expressing various modal properties. S_z is the field intensity, N is a normalisation parameter $S_z = \frac{1}{2}|a|^2\boldsymbol{e} \times \boldsymbol{h}^*\cdot\hat{z}$ and $N = \frac{1}{2}\left|\int_A \boldsymbol{e} \times \boldsymbol{h}^*\cdot\hat{z}\, dA\right|$, η is the fractional power in core, v_g is the group velocity and M_{bm} is the number of bound modes.

Table 2.6 *Weak-guidance expansion of the even HE_{11} mode fields of the step-profile fibre (with superscripts denoting order of expansion).* (Reproduced from Ref. 1)

	Core	Cladding
$\grave{e}_x$	$\dfrac{J_0(\grave{U}R)}{J_0(\grave{U})}$	$\dfrac{K_0(\grave{W}R)}{K_0(\grave{W})}$
$\grave{e}_y$	0	0
$e_z^{(\frac{1}{2})}$	$-2^{\frac{1}{2}}i\dfrac{\grave{U}}{V}\dfrac{J_1(\grave{U}R)}{J_0(\grave{U})}\cos\phi$	$-2^{\frac{1}{2}}i\dfrac{\grave{W}}{V}\dfrac{K_1(\grave{W}R)}{K_0(\grave{W})}\cos\phi$
$\grave{h}_x$	0	0
$\grave{h}_y$	$n_{co}\left(\dfrac{\varepsilon_0}{\mu_0}\right)^{\frac{1}{2}}\dfrac{J_0(\grave{U}R)}{J_0(\grave{U})}$	$n_{co}\left(\dfrac{\varepsilon_0}{\mu_0}\right)^{\frac{1}{2}}\dfrac{K_0(\grave{W}R)}{K_0(\grave{W})}$
$h_z^{(\frac{1}{2})}$	$-2^{\frac{1}{2}}in_{co}\left(\dfrac{\varepsilon_0}{\mu_0}\right)^{\frac{1}{2}}\dfrac{\grave{U}}{V}\dfrac{J_1(\grave{U}R)}{J_0(\grave{U})}\sin\phi$	$-2^{\frac{1}{2}}in_{co}\left(\dfrac{\varepsilon_0}{\mu_0}\right)^{\frac{1}{2}}\dfrac{W}{V}\dfrac{K_1(\grave{W}R)}{K_0(\grave{W})}\sin\phi$
$e_x^{(1)}$	$-\dfrac{\grave{U}U^{(1)}}{J_0(\grave{U})}\left(J_0(\grave{U}R)+\dfrac{2R}{U}J_1(\grave{U}R)+J_2(\grave{U}R)\cos 2\phi\right)$	$\dfrac{\grave{W}W^{(1)}}{K_0(\grave{W})}\left(K_0(\grave{W}R)-\dfrac{2R}{W}K_1(\grave{W}R)-K_2(\grave{W}R)\cos 2\phi\right)$
$e_y^{(1)}$	$-\grave{U}U^{(1)}\dfrac{J_2(\grave{U}R)}{J_0(\grave{U})}\sin 2\phi$	$\grave{W}W^{(1)}\dfrac{K_2(\grave{W}R)}{K_0(\grave{W})}\sin 2\phi$

	Core	Cladding
$e_z^{(3/2)}$	$-\dfrac{2^{3/2} i\grave{U}}{VJ_0(\grave{U})}\left\{\dfrac{\grave{U}^2}{2V^2} J_1(\grave{U} R)+\grave{U}^{(1)}RJ_0(\grave{U} R)\right\}\cos\phi$	$-\dfrac{2^{3/2} i\grave{W}}{VK_0(\grave{W})}\left\{\dfrac{\grave{U}^2}{2V^2} K_1(\grave{W} R)-W^{(1)}RK_0(\grave{W} R)\right\}\cos\phi$
$h_x^{(1)}$	$n_{co}\left(\dfrac{\varepsilon_0}{\mu_0}\right)^{\frac{1}{2}}\left\{\dfrac{\grave{U}^2}{V^2}+\grave{U}\,U^{(1)}\right\}\dfrac{J_2(\grave{U} R)}{J_0(\grave{U})}\sin 2\phi$	$n_{co}\left(\dfrac{\varepsilon_0}{\mu_0}\right)^{\frac{1}{2}}\left\{\dfrac{W^2}{V^2}+\grave{W}W^{(1)}\right\}\dfrac{K_2(\grave{W} R)}{K_0(\grave{W})}\sin 2\phi$
$h_y^{(1)}$	$-\dfrac{n_{co}}{J_0(\grave{U})}\left(\dfrac{\varepsilon_0}{\mu_0}\right)^{\frac{1}{2}}\left\{\dfrac{\grave{U}U^{(1)}}{2}J_0(\grave{U} R)+2U^{(1)}RJ_1(\grave{U} R)\right.$ $\left.+\left(\dfrac{\grave{U}^2}{V^2}+\grave{U}U^{(1)}\right)J_2(\grave{U} R)\cos 2\phi\right\}$	$-\dfrac{n_{co}}{k_0(\grave{W})}\left(\dfrac{\varepsilon_0}{\mu_0}\right)^{\frac{1}{2}}\left\{1-\grave{W}W^{(1)}K_0(\grave{W} R)+2W^{(1)}RK_1(\grave{W} R)\right.$ $\left.+\left(\dfrac{\grave{W}^2}{V^2}+\grave{W}W^{(1)}\right)K_2(\grave{W} R)\cos 2\phi\right\}$
$h_z^{(3/2)}$	$\dfrac{2^{3/2} in_{co}}{J_0(\grave{U})}\left(\dfrac{\varepsilon_0}{\mu_0}\right)^{\frac{1}{2}}\dfrac{\grave{U}U^{(1)}}{V}\left\{\grave{U}J_1(\grave{U} R)-RJ_0(\grave{U}R\right\}\sin\phi$	$-\dfrac{2^{3/2} in_{co}}{K_0(\grave{W})}\left(\dfrac{\varepsilon_0}{\mu_0}\right)^{\frac{1}{2}}\dfrac{\grave{W}W^{(1)}}{V}\left\{\grave{W}K_1(\grave{W} R)-RK_0(\grave{W} R)\right\}\sin\phi$

Eigen value equation	$\grave{U}\dfrac{J_1(\grave{U})}{J_0(\grave{U})}=\grave{W}\dfrac{K(\grave{W})}{K_0(\grave{W})};\ \grave{U}^2+\grave{W}=V^2;\ F_1\cong -1+2\Delta\left\{\grave{U}U^{(1)}+\dfrac{\grave{U}^2}{V^2}\right\};\ F_2\cong -1+2\Delta\grave{U}U^{(1)}$
Modal corrections	$U^{(1)}=\dfrac{\grave{U}\grave{W}}{V^2}\dfrac{K_0(\grave{W})}{K_1(\grave{W})};\ W^{(1)}=-\dfrac{\grave{U}^2}{V^2}\dfrac{K_0(\grave{W})}{K_1(\grave{W})};\ \grave{U}U^{(1)}=-\grave{W}W^{(1)}$

Spread of arrival time of the signal carried by different modes is termed modal dispersion. It can be estimated by computing the difference in transit times of all the modes present if the signal is carried without mode conversion. Otherwise an average transit time spread would occur.

When only the jth mode is involved (usually when the waveguide is operating in a single-mode situation), the intramodal dispersion is due to the frequency spread of the carrier.

Assuming the refractive index of material is not changing with frequency

$$\delta t_j = z\frac{d^2\beta_j}{d\omega^2}\,\delta\omega = \frac{z}{2\pi c}\left\{\lambda^2\frac{d^2\beta_j}{d\lambda^2} + 2\lambda\frac{d\beta_j}{d\lambda}\right\}\delta\lambda \tag{2.68}$$

If the refractive index of the material is changing with λ, the intramode waveguide dispersion is influenced by the refractive-index change. No convenient expression can be derived. However, for plane-wave propagation the material dispersion can be expressed as

$$\delta t_g = -\frac{z}{c}\lambda\frac{d^2n}{d\lambda^2}\,\delta\lambda \tag{2.69}$$

This expression is derived from the plane-wave group velocity through a wavelength-dependent index region. This is an empirical wavelength-dependence relationship which most dielectric materials obey:

$$v_g = c\left(n(\lambda) - \lambda\frac{dn(\lambda)}{d\lambda}\right)^{-1} \tag{2.70}$$

2.4 Cut off

For each mode β_j is real when the waveguide is lossless. Hence, the range of values for β_j for bound modes is when

$$n_{cl}k < \beta_j < n_{co}k$$

i.e. β must be between c/n_{co} and c/n_{cl} since $v_p = \omega/\beta$

The cut off corresponds to when

$$\beta_j = n_{cl}k \text{ or } W = (n_{cl}^2k^2 - \beta_j^2)^{\frac{1}{2}} = 0 \tag{2.71}$$

i.e. when the propagation constant within the waveguide is indistinguishable from that of the plane wave in the cladding region. In other words, the wave is radiated as a plane wave.

At cut off

$$u_j = V = [(n_{cl}^2 - n_{co}^2)k^2]^{\frac{1}{2}} \tag{2.72}$$

2.5 Weakly guiding fibre

This nomenclature is given to a fibre with n_{cl} differing by a small amount from n_{co} such that Δ is of the order of 0·01. Even with small Δ the guidance of a given mode is still good, as measured by the fraction of power within the core, which can still be almost 100%, and as measured by its sensitivity to bending, which can be low. In fact, guidance is a function of operating value V such that the modes far from cut off at this V value will be strongly guided while the mode nearest cut off will only be weakly guided. Hence, the name weakly guided fibre must be interpreted carefully.

When $n_{cl} \sim n_{co}$ the propagation constant β lies between kn_{cl} and kn_{co} and hence

$$\beta \doteqdot kn_{co} \doteqdot kn_{cl} \tag{2.73}$$

The modes must resemble plane-wave propagation with $e_z \sim 0$ and $h_z \sim 0$. Hence, the plane-wave relationship of $\boldsymbol{H}_t = (\varepsilon_0/\mu_0)^{\frac{1}{2}} n_{co}\hat{z} \times \boldsymbol{E}_t$ applies. The existence of index difference introduces a small amount of polarisation dependence since reflection with $\boldsymbol{E}$ parallel or $\boldsymbol{H}$ parallel are different. The weakly guided modes for fibres with index variations in the transverse direction can be constructed for spatial dependence by solving the scalar wave equation assuming negligible polarisation dependence.

Take the fibre with circular core whose refractive index is a function of r:

$$n^2(R) = n_{co}^2\,[1 - 2\Delta f(R)] \tag{2.74}$$

where

$$R = r/\rho$$

n_{co} is the maximum refractive index and ρ is the core radius or a scaling length, $\Delta = (n_{co}{}^2 - n_{cl}{}^2)/2n_{co}{}^2$ is small and $f(R) = 0$ at minimum index and $f(R) > 1$ at the maximum index.

$$\boldsymbol{E} = \boldsymbol{E}_t \exp\{i(\tilde{\beta} + \delta\tilde{\beta})z\}$$

$$\boldsymbol{H}_t = n_{co}\left(\frac{\varepsilon_0}{\mu_0}\right)^{\frac{1}{2}} \hat{z} \times \boldsymbol{E}_t$$

$\tilde{\beta}$ is the approximate phase constant, and $\delta\tilde{\beta}$ is a correction term to account for the polarisation dependence. $\boldsymbol{E}_t$ can be found by solving the scalar wave equation since for small Δ and the $\nabla_t \ln n^2$ term can be ignored.

$$\{\nabla_t^2 + k^2 n^2(r,\phi) - \tilde{\beta}^2\}\,\psi = 0 \tag{2.75}$$

where ψ is either e_x or e_y with $\boldsymbol{E}_t = e_x\hat{x} + e_y\hat{y}$ or $e_r\hat{r} + e_\phi\hat{\phi}$

In normalised cylindrical co-ordinate system

$\psi = F_l\,(R) \sin l_\phi$ or $F_l\,(\mathrm{R}) \cos l_\phi$ with $F_l\,(R)$ satisfying

$$\left\{\frac{d^2}{dR^2} + \frac{1}{R}\frac{d}{dR} + \frac{e^2}{R^2} + \tilde{u}^2 - V^2 f(R)\right\} F_l(R) = 0 \tag{2.76}$$

For step profile,

$n(R) = n_{co}$ or $f = 0 \quad 0 \leqslant \mathrm{R} < 1$
$n(R) = n_{cl}$ or $f = 1 \quad 1 < \mathrm{R} < \infty$

For $l = 0$

$$\left\{\frac{d^2}{dR^2} + \frac{1}{R}\frac{d}{dR} + \tilde{u}^2\right\} F_0(R) = 0 \tag{2.77a}$$

$$\text{for } 0 \leqslant R < 1 \quad F_0(R) = J_0(\tilde{u}R)/J_0(\tilde{u}) \tag{2.78a}$$

and normalised so that $F_0 = 1$ at $R = 1$

$$\left\{\frac{d^2}{dR^2} + \frac{1}{R}\frac{d}{dR} + \tilde{u}^2 - V^2\right\} F_0(R) = 0 \tag{2.77b}$$

$$\text{for } 1 < R < \infty \quad F_0(R) = K_0(\tilde{\omega}R)/K_0(\tilde{\omega}) \tag{2.78b}$$

The eigenvalue equation can be derived by differentiating eqns. 2.78*a* and *b* and equating, using the fact that $F_0(R)$ is continuous so that $F_0(R)$ of the two regions at $R = 1$ are equal. This is the same as matching the tangential field components. Using recurrent relationships of the Bessel functions, the eigenvalue equation is

$$\tilde{u}\frac{J_1(\tilde{u})}{J_0(\tilde{u})} = \tilde{\omega}\frac{K_1(\tilde{\omega})}{K_0(\tilde{\omega})} \tag{2.79}$$

Note $\dfrac{dJ_0(z)}{dz} = -J_1(z)$ and $\dfrac{dK_0(z)}{dz} = -K_1(z)$

This is somewhat simpler than the exact eigenvalue equation, but it can still only be solved by numerical techniques. The β value obtained, in general, must be corrected by the $\delta\beta$ term, which is derived through a perturbation analysis. However, for HE_{1m} modes with $l = 0$, $\delta\beta$ can be shown to be negligibly small. These modes are given the label LP_{0m} modes.

The expressions for the field components for the LP_{01} mode for x and y are therefore in directions $\phi = 0$ and $90°$, respectively:

$$e_x = F_0(R)\exp(i\tilde{\beta}z) \quad e_y = F_0(R)$$

or

$$h_y = n_{co}\left(\frac{\varepsilon_0}{\mu_0}\right)^{\frac{1}{2}} F_0(R)\exp(i\tilde{\beta}z) \quad h_x = n_{co}\left(\frac{\varepsilon_0}{\mu_0}\right)^{\frac{1}{2}} F_0(R)$$

with small components of e_z and h_z which can be obtained in principle from eqns. 2.23 and 2.28, but is negligible for $l = 0$ modes; for $l \neq 0$ modes more accurate estimations are needed.
Fibre with step-index core has been solved by assuming the core to be a cylinder in an infinite cladding.

Case 1: classical approach: The wave equation is reduced to a scalar wave equation which is solved for the particular boundary condition.

Case 2: A. W. Snyder et al approach[1]*:* The formulation starts by assuming the core index not to be uniform, but dependent on the transverse co-ordinates. The modal description and model field expressions allow waveguide characteristics to be examined in detail.

Case 3: By assuming Δ to be small, simplification of the expressions results. This helps to reduce computational efforts, gains clearer physical understanding, and allows explicit expressions for fibres with varying index core to be computed.

These analyses permit a systematic description to be given to the fibre characteristics in terms of waveguide modes. Each waveguide operating at a particular V value will have a finite set of propagating or guided modes and a continuum of radiative field. Each of the modes and the radiative field has prescribed propagation constants and field distribution for its field components. The jth propagating mode is characterised, besides its propagation constant β_j, by

$$\frac{\mu_j}{\rho} = (k^2n_{co}^2 - \beta_j^2)^{\frac{1}{2}} \text{ and } \frac{\omega_j}{\rho} = (\beta_j^2 - k^2n_{cl}^2)^{\frac{1}{2}}$$

which are the transverse wave numbers. Each propagating mode has a cut off which occurs when $W_j = 0$. The signal velocity carried by the jth mode is given by the group velocity $d\omega/d\beta$ associated with the mode, and the power in the core is expressible as a fraction of the total power of the mode.

A fibre supporting only the lowest-order mode at the operating wavelength is referred to as a single-mode fibre, while a fibre supporting more than one mode at the operating wavelength is called a multimode fibre. When the core is a step index the fibre is a step-index fibre, while a fibre with a core whose index distribution follows a power law, particularly parabolic power in the radial direction, and is circularly symmetric, is called a graded-index fibre. Of course, it is possible to have a graded-index single-mode fibre.

The bandwidth of the fibre is caused by the change in group velocity. It is also referred to as dispersion or time spread of the arrival of an impulse function. For multimode fibre, dispersion is introduced predominantly through the group-velocity differences of the modes, and is referred to as modal dispersion. When operating with a broad spectral-width light source the dispersion could be dominated by material dispersion, which is of the order of a few picoseconds over a distance of 1 km for a light source with 1 nm spectral width.

For single-mode fibre, dispersion is the combined effect of material dispersion and waveguide dispersions when operating with a light source of certain spectral width. If the light source is strictly monochromatic, i.e. a single-frequency source, dispersion is dependent on the signal bandwidth only.

2.6 Reference

1 SNYDER, A. W., and LOVE, J. D.: 'Optical waveguide theory, (Chapman & Hall, 1983)

Chapter 3

Physical-optics approach

3.1 Introduction

While electromagnetic-wave solutions offer an in-depth and accurate description of the fibre characteristics, physical optics, and especially geometrical optics, provide a readily visualisable description which is helpful towards developing a physical understanding of the performance of the fibre. It provides design guidelines as well as suggesting how electromagnetic solutions can be constructed by suitable approximations. In the course of study of geometrical-optics applications to fibres, particularly multimode fibres, the validity of the ray approximation can be demonstrated by comparing the results with those obtained from the exact electromagnetic-wave solution. This has enabled the transition between ray optics and electromagenetic wave to be meaningfully portrayed.

Consider the concept of an optical beam or a ray in relation to the infinite plane wave. If a plane wavefront is infinite in extent, it satisfies Maxwell's equations and can be exactly characterised by a direction of propagation normal to the wavefront and a polarisation direction of the tranverse E vector. When a cylindrical optical beam has a cross-section measured in many wavelengths across its diameter, the beam spread is small, especially over short distances. It is reasonable to expect that the beam front behaves as a localised plane wave of finite extent. This approximation is theoretically difficult to handle, and has not been satisfactorily quantified in a general and useful way. Nevertheless, a slender beam of light of 0·1 mm in diameter at 1 μ wavelength only diverges at approximately 1/100th radian, so that, when propagating to a point 100 wavelengths away, the diameter only increased by 1%. Its behaviour is expected to resemble closely that of a plane wave.

In a straight fibre the ray approximation seemingly has good validity. This can be explained readily in a planar waveguide. The propagating modes of a planar waveguide can be shown to be equivalent to the intersection of two plane waves. Hence, a ray analysis is exact. In the fibre the propagating of modes can be shown to be able to be constructed from a spectrum of plane waves, indicating again that ray representation should be exact.

We start by examining the plane-wave characteristics. The solution to the boundary value problem of a plane-wave incident at an interface can be found in any book on optical or electromagnetic-wave theory and will not be repeated here. The important conclusions are:

(i) *Law of reflection*

$$\theta = \theta'' \tag{3.1}$$

Angles are relative to (x, z) plane.
The y plane is an infinite boundary separating regions 1 and 2 with refractive indices n_1 and n_2, respectively.
Incident wave $e^{j\beta_1 (y \sin\theta - z \cos\theta)}$.
Refracted wave $e^{j\beta_2 (y \sin\theta' - z \cos\theta')}$.
Reflected wave $e^{j\beta_1 (y \sin\theta'' - z \cos\theta'')}$.

(ii) *Law of refraction (Snell's law)*

$$\frac{\sin\theta}{\sin\theta'} = \frac{n_2}{n_1} \tag{3.2}$$

(iii) *Power reflection coefficient for E normal to plane of incidence or TM case*

$$R_E = \frac{(n_1 \cos\theta - \sqrt{n_2{}^2 - n_1{}^2 \sin^2\theta})^2}{(n_1 \cos\theta + \sqrt{n_2{}^2 - n_1{}^2 \sin^2\theta})^2} \tag{3.3}$$

(iv) *Power-transmission coefficient*

$$T_E = \frac{4n_1 \cos\theta \sqrt{n_2{}^2 - n_1{}^2 \sin^2\theta}}{(n_1 \cos\theta + \sqrt{n_2{}^2 - n_1{}^2 \sin^2\theta})^2} \tag{3.4}$$

(v) *Power conservation*

$$R_E + T_E = 1$$

(vi) *Total internal reflection*

If $n_1 > n_2$ for $\sin\theta = n_2/n_1$

$$T_E = 0$$

$$R_E = 1$$

θ is referred to as θ_c, the critical angle. As $\theta > \theta_c$, R_E is complex but $|R_E| = 1$,

T_E is purely imaginary. The refracted field propagates even after $\theta > \theta_c$. The propagation is described by

$$e^{j\beta_2(y \sin\theta' - z\cos\theta')}$$
$$= e^{j\beta_2\left(y\frac{n_1 \sin\theta}{n_2} + jz\sqrt{\left(\frac{n_1}{n_2}\right)^2 \sin^2\theta - 1}\right)}$$
$$= e^{j\beta_2 y\frac{n_1 \sin\theta}{n_2}}\, e^{-\beta_2 z\sqrt{\left(\frac{n_1}{n_2}\right)^2 \sin^2\theta - 1}} \tag{3.5}$$

The field has a phase constant change along y, but decays exponentially along z with increasing rate as θ is increased. The reflected field suffers an additional phase shift. This phase shift can be interpreted as a movement of the actual point when the wave is reflected. Hence the reflection can schematically be shown as Fig. 3.1.

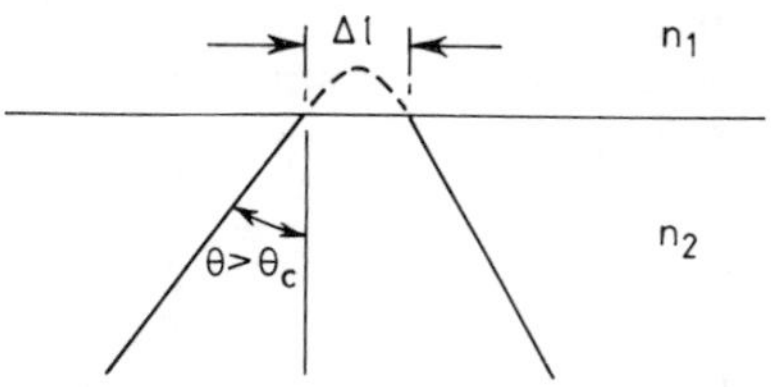

Fig. 3.1 *Schematic of reflection*

Δl is called the Goos–Haenchen shift and can be demonstrated through power-flow considerations to equal

$$\Delta l = \frac{2}{kn_2\,\theta}\;\frac{1}{(\theta_c^2 - \theta^2)^{\frac{1}{2}}} \tag{3.6}$$

which is very small except when $\theta \approx \theta_c$

(vii) *Brewster angle*
For the TE wave

$$R_H = \frac{\left(n_1\cos\theta - \frac{n_1}{n_2}\sqrt{n_2^2 - n_1^2\sin^2\theta}\right)^2}{\left(n_2\cos\theta - \frac{n_1}{n_2}\sqrt{n_2^2 - n_1^2\sin^2\theta}\right)^2} \tag{3.7}$$

$$T_H = \frac{4n_1\cos\theta\sqrt{n_2^2 - n_1^2\sin^2\theta}}{\left(n_2\cos\theta - \frac{n_1}{n_2}\sqrt{n_2^2 - n_1^2\sin^2\theta}\right)^2}$$

$$R_H + T_H = 1$$

At $\sin\theta_B = n_2/\sqrt{n_1^2 + n_2^2}$

$R_H = 0$. This angle is as the Brewster angle when the plane wave is totally transmitted.

3.2 Planar-waveguide ray description

Equivalent plane waves in a planar waveguide can be generated from the waveguide solution. In a planar waveguide, the wave propagates with phase constant β in the z direction since

$$e_y = \frac{\cos ux}{\cos u} = \frac{e^{iux} + e^{-iux}}{2 \cos u} = e_y{}^+ + e_y{}^-$$

e_y can be considered as made up of two components $e_y{}^+$ and $e_y{}^-$ propagating in direction z with $e^{i\beta z}$, and in direction x with e^{+iux}, u is the tranverse propagation constant.

Let the direction cosine of a pair of plane-wave vectors k^+ be

$$k^{\pm} = (\pm u/\rho, 0, \beta) = kn_2 (\pm \sin \theta_2, 0, \cos \theta_z)$$

This pair of plane waves propagate in direction x with propagation constant $\pm u$ (Fig. 3.2). Hence, $e_y{}^+$ and $e_y{}^-$ can be interpreted as the electric-field components of two intersecting plane waves propagating with angle θ_x relative to the axis of propagation.

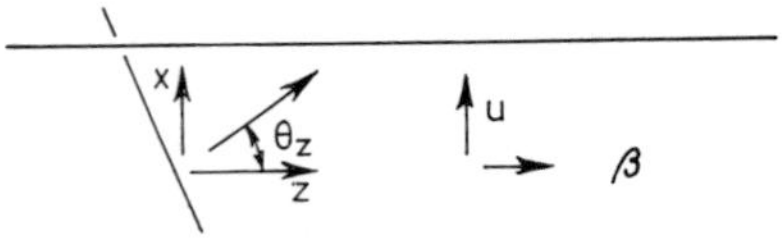

Fig. 3.2 *Pair of plane waves propagating in direction x with propagation constant ±u*

The ray description is equivalent to a plane-wave description. In a waveguide the propagating modes can be considered as a spectrum of intersecting plane waves so arranged that the resulting wave satisfies the boundary conditions. This description is easily demonstrated for a planar waveguide. In any case, the plane-wave description does not necessarily provide a clearer physical insight, particularly when a dense spectrum of intersecting plane waves is involved. However, the equivalence of ray tracing to plane-wave description, and in turn, linking to the waveguide solution, lends credence to applying geometrical-ray techniques to solve particular multimode problems.

The ray description can be used to give a physical description of dispersion through calculating the length of ray paths as waveguide modes represented by rays propagated at different angles down the waveguide.

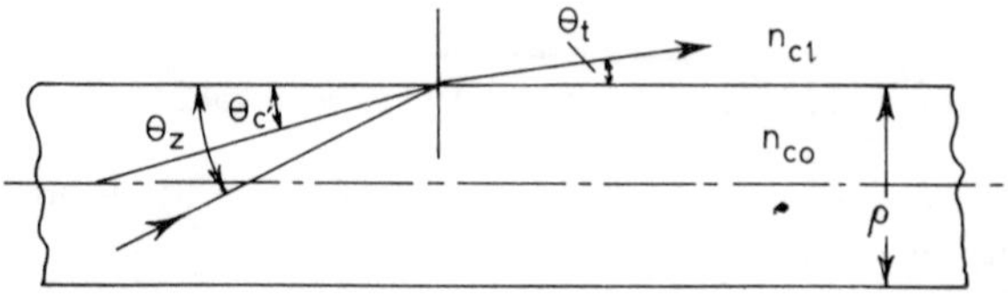

Fig. 3.3 *Physical description of dispersion*

Let $\theta_{c'}$ be the complement of the critical angle θ_c (Fig. 3.3):

$$\theta_{c'} = \cos^{-1}(n_{cl}/n_{co}) = \sin^{-1}[1 - (n_{cl}/n_{co})^2]^{\frac{1}{2}} \quad (3.9)$$

Snell's law gives

$$n_{co} \cos \theta_z = n_{cl} \cos \theta_t \text{ if } \theta_z > \theta_c$$

otherwise it is totally internally reflected.

It can be seen that the distance L_p between two reflections for a planar uniform index waveguide of thickness 2ρ is

$$L_p = \frac{2\rho}{\sin \theta_z} \equiv \frac{2\rho n_{co}}{(n_{co}^2 - \bar{\beta}^2)^{\frac{1}{2}}} \equiv \frac{L_o}{n_{co}} \quad (3.10)$$

where $\beta = n_{co} \cos \theta_z$ and L_o = optical path length. Hence, $L_p \cos \theta_z$ gives the distance along the guide between the successive reflections:

$$L_p \cos \theta_z = z_p = \frac{2\rho\bar{\beta}}{(n_{co}^2 - \bar{\beta}^2)^{\frac{1}{2}}} \quad (3.11)$$

The transit time for covering distance z along the waveguide is

$$t = \frac{z}{z_p} \frac{L_p}{v_g} = \frac{z\, n_{co}^2}{c\bar{\beta}} = \frac{z}{c} \frac{n_{co}}{\cos \theta_z} \quad (3.12)$$

v_g = plane wave velocity = c/n_{co}

In a region with refractive-index variations, the ray trajectory is governed by the ray eikonal equation which describes the energy-flow direction (Fig. 3.4):

$$\frac{d}{ds}\left\{n(\mathbf{r})\frac{d\mathbf{r}}{ds}\right\} = \nabla n$$

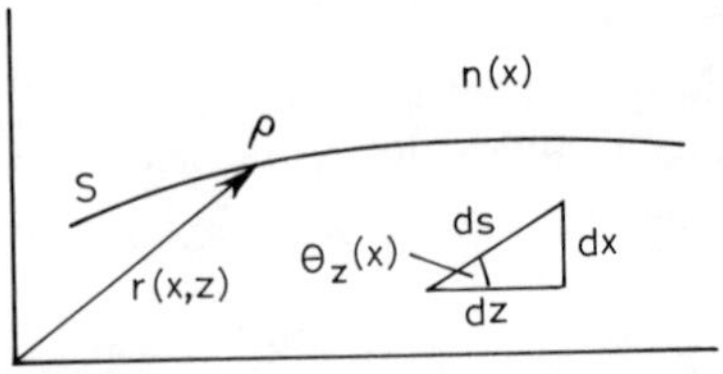

Fig. 3.4 *Ray trajectory in region with refractive-index variations*

with co-ordinates as shown. The ray path in a graded-index region can be derived as follows. We have

$$\frac{d}{ds}\left\{n(x)\frac{dx}{ds}\right\} = \frac{d\,n(x)}{dx} \quad \text{in direction } x \tag{3.14}$$

and

$$\frac{d}{ds}\left\{n(x)\frac{dx}{ds}\right\} = 0 \quad \text{in direction } z \tag{3.15}$$

Introduce

$$\frac{dx}{ds} = \sin\theta_z(x) \text{ and } \frac{dz}{ds} = \cos\theta_z(x)$$

Using eqn. 3.15

$$\frac{d}{ds}\left\{n(x)\frac{dx}{ds}\right\} = 0 \text{ becomes } n(x)\cos\theta_z(x) = \text{constant} = n(0)\cos\theta_z \text{ at } x = 0$$

Therefore $n(x)\cos\theta_z(x) = n(0)\cos\theta_z(0)$ (3.16)

The ray path is fixed by the initial launching condition, and if $n(x)$ decreases as x increases, the ray will turn back at $x = x_c$

$$n(x_c) = n(0)\cos\theta_z(0) \text{ when } \cos\theta_z(x_c) = 1 \text{ i.e. } \theta_z(x_c) = 0°$$

Hence the ray path schematically is as shown in Figs. 3.5*a* and *b*.

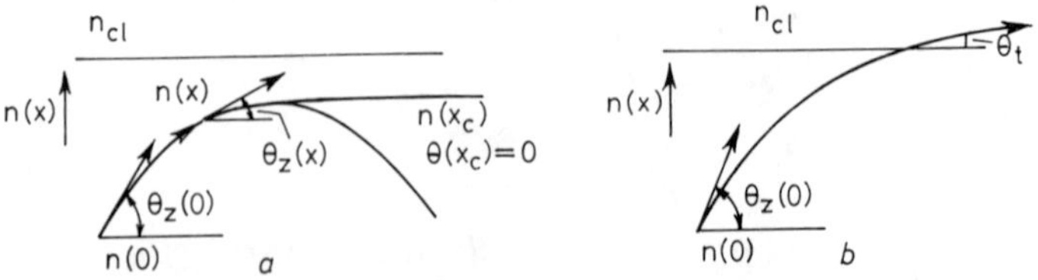

Fig. 3.5 *Schematic of ray path*

$$\bar{\beta} = n(x)\cos\theta_z(x) = n(x)\,dz/ds \equiv n(x_c) \tag{3.17}$$

at the ray turning point.

Using eqn. 3.14

$$\frac{d}{ds}\left\{n(x)\frac{dx}{ds}\right\} = \frac{dn(x)}{dx}$$

and $\quad \bar{\beta} = n(x)\dfrac{dz}{ds}$ from eqn. 3.17

$$\frac{dz}{ds}\frac{d}{dz}\left\{n(x)\frac{dz}{ds}\frac{dx}{dz}\right\} = \frac{d\,n(x)}{d\,x}$$

Therefore

$$\frac{\bar{\beta}}{n(x)}\frac{d}{dz}\left\{n(x)\frac{\bar{\beta}}{n(x)}\frac{dx}{dz}\right\} = \frac{d\,n(x)}{dx} \quad \text{and} \quad \bar{\beta}^2\frac{dx^2}{dz^2} = \frac{1}{2}\frac{d\,n^2(x)}{dx}$$

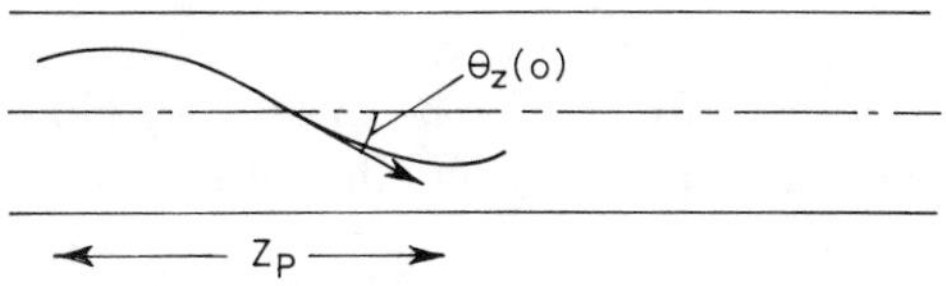

Fig. 3.6 *Distance between two turning points*

Setting

$$\frac{d^2x}{dz^2} = \frac{dx}{dz}\frac{d\left(\dfrac{dx}{dz}\right)}{dx} = x'\frac{dx'}{dx}$$

$$\bar{\beta}^2\frac{x'dx'}{dx} = \frac{1}{2}\frac{d\,n^2(x)}{dx}$$

Integrating $\bar{\beta}^2\, x'dx' = \frac{1}{2}\, d\, n^2(x)$ gives $\bar{\beta}\, x'^2 = n^2(x) + A$

since $x' = 0$ and $n(x) = \bar{\beta}$ at $x = x_c$, $A = n^2 = \bar{\beta}^2$

Therefore

$$\bar{\beta}\frac{dx}{dz} = \left\{n^2(x) - \bar{\beta}^2\right\}^{\frac{1}{2}}$$

Hence

$$z(x) = \bar{\beta}\int_0^x \frac{dx}{(n^2(x) - \bar{\beta}^2)^{\frac{1}{2}}} \tag{3.18}$$

if $z = 0$ when $x = 0$, i.e. constant of integration $= 0$. The distance between two turning points z_p (Fig. 3.6) is obtained by taking the integral between $\pm x_c$.

Therefore

$$z(p) = \bar{\beta} \int_{-x_c}^{x_c} \frac{dx}{(n^2(x) - \bar{\beta}^2)^{\frac{1}{2}}} \tag{3.19}$$

$$L_o = \int_P^Q n(x)\, ds \text{ can be shown to equal to} \int_{-x_c}^{x_c} \frac{n^2(x)\, dx}{(n^2(x) - \bar{\beta}^2)^{\frac{1}{2}}} \tag{3.20}$$

3.3 Ray transit time

Along a graded refractive-index region the transit time for the ray is determined by integrating the incremental time as the ray travels at speed c/n.

$$t = \int \frac{n(x)}{c}\, ds = \frac{1}{c\bar{\beta}} \int n^2(x) dz \tag{3.21}$$

This integral is not necessarily able to be evaluated. However, if an optical path length L_o for a ray of half-period length z_p is known, then for transit time over distance z,

$$t = \frac{z}{z_p} \frac{L_o}{c}$$

For a graded region with a parabolic profile let

$$n^2(x) = n_{co}^2 \{1 - 2\Delta(x/\rho)^2\} \quad \text{(Fig. 3.7)}$$

Since

$$n(x_c) = \bar{\beta} = n_{co} \cos \theta_z(0)$$

$$= n_{co} \left(1 - 2\Delta \left(\frac{x_c}{\rho}\right)^2\right)^{\frac{1}{2}}$$

$$x_c = \pm \rho \frac{\sin \theta_z(0)}{\sqrt{2\Delta}}$$

$$= \pm \rho \frac{n_{co}^2 [1 - \cos^2\theta_z(0)]}{n_{co}^2 \sqrt{2\Delta}}$$

$$= \pm \rho \frac{n_{co}^2 - \bar{\beta}^2}{n_{co}^2 \sqrt{2\Delta}}$$

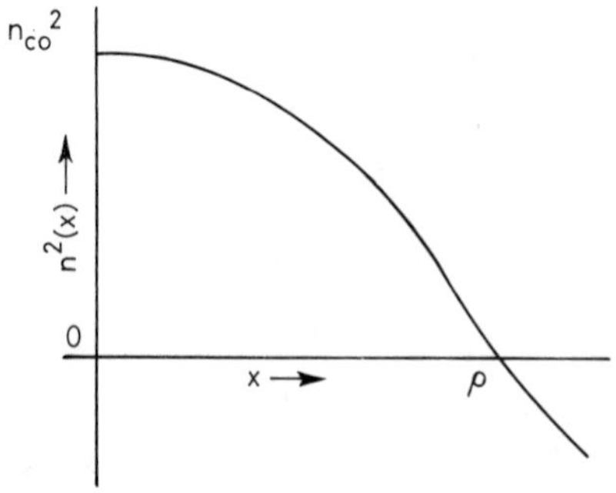

Fig. 3.7 *Graded region with parabolic profile*

Substitute into eqn. 3.18

$$z(x) = \bar{\beta} \int_0^x \frac{dx}{(n^2(x) - \bar{\beta}^2)^{\frac{1}{2}}} \tag{3.22}$$

It can be shown that

$$z_p = \frac{\pi \rho \bar{\beta}}{n_{co} \sqrt{2\Delta}} \tag{3.23}$$

and

$$L_o = z_p \frac{n_{co}^2 - \bar{\beta}^2}{2\bar{\beta}} \tag{3.24}$$

Hence, the ray transit time

$$t = \frac{z(\bar{\beta}^2 + n_{co}^2)}{2c\bar{\beta}} = \frac{z}{c} n_{co} \frac{\left(\cos\theta_z + \dfrac{1}{\cos\theta_z}\right)}{2}$$

$$= \frac{z}{c} n_{co} \,(\text{terms of } \theta^4 \text{ and higher}) \tag{3.25}$$

This result shows that transit time is less dependent on θ_z than the step-index case. For a hyperbolic secant profile

$$n^2(x) = n_{co}^2 \operatorname{sech}^2\left(\sqrt{2\Delta}\frac{x}{\rho}\right)$$

$$x_c = \frac{\rho}{\sqrt{2\Delta}} \cosh^{-1}\left(\frac{n_{co}}{\beta}\right)$$

Leading to

$$z_p = \pi\rho/\sqrt{2\Delta} \text{ and } L_o = n_{co}\, z_p$$

Therefore

$$t = \frac{zn_{co}}{c} \tag{3.26}$$

which is the same as the axial ray, indicating zero dispersion.

3.4 Circular-cross-section fibre

In a cylindrical fibre with a constant circular cross-section, the ray approach offers a convenient physical insight into what may be the expected characteristics. Several ray paths are relevant in a fibre. The meridional rays are those rays which cross the fibre axis between reflections in the straight waveguide. The skew rays are those which never cross the fibre axis.

A meridian ray lies in a plane $\Phi = \Phi_i$. The angle of rays within that plane makes angle θ_z with the z direction. A skew ray follows a helical path and touches a cylinder of radius r_{ic} as a caustic (Fig. 3.8).

It has a direction θ_z which indicates its inclination relative to the z axis. A second angle is needed to indicate the skewness. This angle is defined as the angle in the core cross-section between the tangent to the interface and the projection of the ray path in the (r, Φ) plane (Fig. 3.9).

At point P if $\theta_z < \theta_{c'}$ total internal reflection will take place. This leads some skew rays to be bounded, refracting or leaky (tunnelling). This last type is when $\theta_c \leqslant \theta_z \leqslant \pi/2$ and $\alpha_{c'} \leqslant \alpha \leqslant \pi/2$.

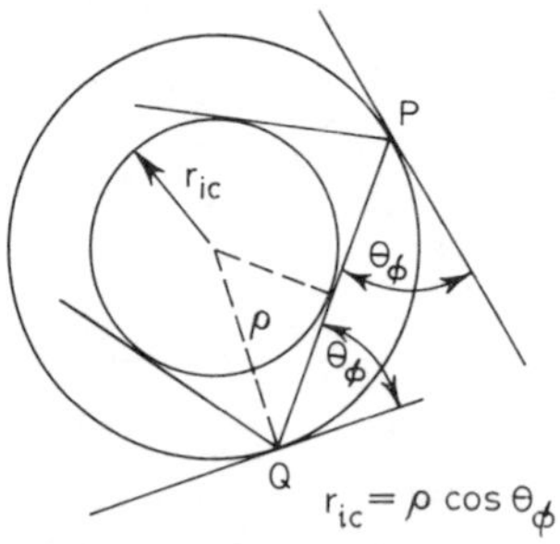

Fig. 3.8 *Skew ray follows helical path*

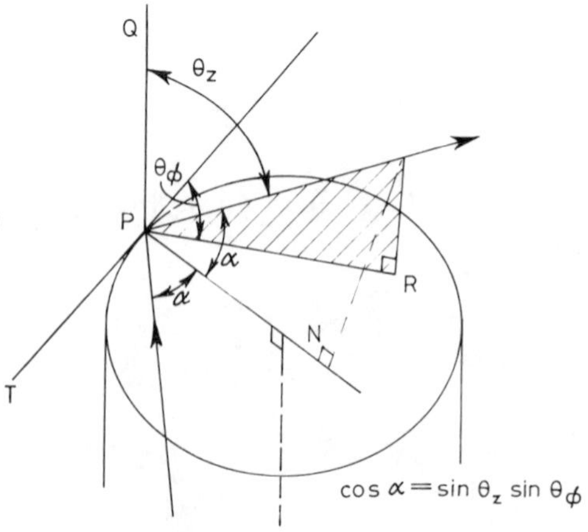

Fig. 3.9 *Second angle to indicate skewness*

Ray transit time for the fibre can be deduced as in the case of the planar guide. In a step-index fibre the path length, as expected, is dependent on the angle θ_z only and is independent whether the ray is skewed or meridional.

$$t = \frac{z}{c} \frac{n_{co}}{\cos \theta_z} \tag{3.27}$$

For a graded index the skew rays obviously propagate at different velocities for different θ_x. Hence, the transit time will not be dependent on θ_z only. In general, graded profile can equalise transit time for different meridional rays just as in the planar waveguide case, but cannot correct for skew rays at the same time.

The ray approach can provide some insight into pulse spreading, launching efficiency and some problems concerning discontinuity and variations in the z direction. This discussion of ray propagation ends with a look at the pulse-spreading problem. The other aspects are to be treated in Chapter 4.

3.5 Pulse spreading and leaky-wave effects

Pulse spread in a ray description is deduced by computing the difference of transit time between the various bounded rays and tunnelling or leaky rays. Most tunnelling rays die away rapidly and contribute little to the pulse-spread effect. However, if dispersion is estimated by measurement made over a relatively short length of fibre, the tunnelling rays can give misleading information.

The tunnelling rays in a graded-index fibre can be shown to occur when $\bar{\beta} < n_{cl}$ and $n_{cl}^2 < \bar{\beta}^2 + \bar{l}^2$

where

$$\bar{\beta} = n(r) \cos \theta_z(r) \tag{3.28}$$

and

$$\bar{l} = \frac{r}{\rho} n(r) \sin \theta_z(r) \cos \theta_\phi(r) \tag{3.29}$$

By eliminating $\theta_z(r)$ from $\bar{l}$ and $\bar{\beta}$ in eqns. 3.28 and 3.29

$$\cos \theta_\phi(r) = \frac{\rho}{r} \frac{\bar{l}}{(n^2(r) - \bar{\beta}^2)^{\frac{1}{2}}}$$

When

$$\theta_\phi(r) = 0$$

$$g(r) \equiv n^2(r) - \bar{\beta}^2 - \bar{l}^2 \frac{\rho^2}{r^2} = 0 \tag{3.30}$$

giving the two solutions for the core region of r corresponding to r_{ic} and r_{tp} (turning point).

For tunnelling rays $g(r) < 0$ until r is beyond a certain radius in the cladding. We call this radius

$$r_{rad} = \rho \bar{l}/(n_{ce}^2 - \bar{\beta}^2)^{\frac{1}{2}}$$

given by

$$g(r_{rad}) = 0 \tag{3.31}$$

Hence the tunnelling ray radiates by tunnelling through into the cladding at $r > r_{rad}$.

The transmission coefficient for tunnelling rays can be computed from the scalar wave equation (see Ref. 1 of Chap. 2).

The form of the transmission coefficient is $T \alpha\, e^{-2ky}$

where

$$y = \int_{r_{tp}}^{r_{rad}} \left\{ \bar{\beta}^2 + \bar{l}^2 \frac{\rho^2}{r^2} - n^2(r) \right\}^{\frac{1}{2}} \tag{3.32}$$

Hence T decreases rapidly as $(r_{rad} - r_{tp})$ increases, indicating that tunnelling rays can remain within the fibre with significant power over long distances.

3.6 Dispersion

For a planar guide with step-index core

$$t_{min} = \frac{z}{c} n_{co} \qquad t_{max} = \frac{z}{c} \frac{n_{co}}{\cos \theta_c}$$

$$t_d = t_{max} - t_{min} = \frac{z}{c} n_{co} \left\{ \frac{n_{co}}{n_{cl}} - 1 \right\}$$

For Δ to be small

$$t_d = z\, n_{co}\, \Delta/c \risingdotseq z\, n_{co}\, \theta_c^2/2c \tag{3.33}$$

For cylindrical fibre with graded-index core the dispersion due to paraxial rays can readily be estimated by using the planar-guide analysis. If the graded-index profile follows a power law

$$n^2(x) = n_{co}^2 \left(1 - 2\Delta \left| \frac{x}{\rho} \right|^q \right) \tag{3.34}$$

$$t(\bar{\beta}) = \frac{z}{c} \frac{n_{co}}{q+2} \left(q \frac{n_{co}}{\bar{\beta}} + 2 \frac{\bar{\beta}}{n_{co}} \right) \tag{3.35}$$

A graphical plot of $t(\bar{\beta})$ versus $\bar{\beta}$ is given in Fig. 3.10, t_{max} and t_{min} for bounded rays depends on the range of β values and q.

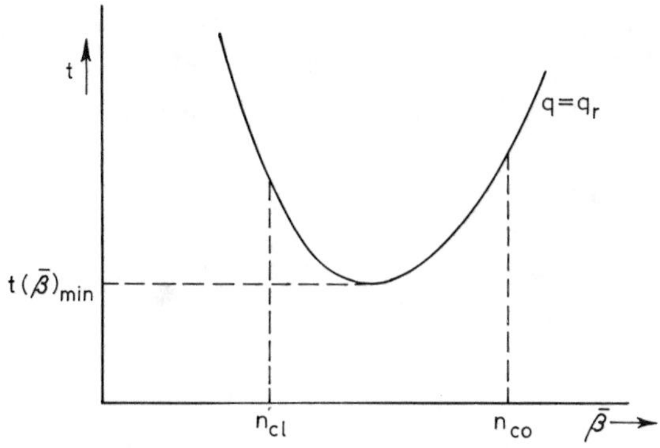

Fig. 3.10 *Plot of t ($\bar{\beta}$) versus β*

For $q > 2$

$$t_d = t(n_{cl}) - t(n_{co})$$

For $0 < g \leqslant 2 - 4\Delta$

$$t_d = t(n_{co}) - t(n_{cl})$$

For $2 - 4\Delta \leqslant g \leqslant 2$

$$t_d = t(n_{cl}) - t(\bar{\beta})_{min} \text{ or } t(n_{co}) - t(\bar{\beta})_{min} \tag{3.36}$$

whichever is the larger.

The minimum pulse spread can be shown from eqn 3.36 to be when q is such that $t(n_{cl}) - t(\beta)_{min}$ equals $t(n_{co}) - t(\beta)_{min}$. The minimum pulse spread will then be

$$t_d \doteqdot \frac{z}{c}\frac{\Delta^2}{8}\, n_{co}.$$

This occurs for q nearly equal to 2. Hence t_d is smaller by a factor of $\Delta^2/8$ over that of a step-index fibre. When bounded skew rays are included, the index-profile optimisation is much more complex, and if tunnelling rays are still significant, the dispersion will be length dependent as well. The region in which tunnelling rays radiate significantly can be estimated. This region is sometimes referred to as the spatial transient region, as opposed to spatial steady state when leakage is taking place.

Even when excited with a source which has a broad radiation lobe, the refracted rays die away very rapidly within the immediate vicinity of the source. The main contribution of the spatial transient is due to the tunnelling rays. The power in the core within this region will have contributions from both bounded rays and tunnelling rays. The power in the tunnelling rays depends on the distribution of power in the angular spectrum of the light source. For a diffused source the power is nearly equal at the beginning of the fibre. The impulse response is modified by the tunnelling rays, which add a tail to the pulse. The duration of the spatial transient is dependent on the fibre operating parameter V. It is possible to show that after

$$z = \frac{\rho}{2\theta c} \exp\,(V/2) \tag{3.37}$$

tunnelling-ray power has decreased to 10% of the starting value. For $V = 30$ $\rho = 25\ \mu$m and $\theta_c = 0{\cdot}14$, $z = 300$m.

Chapter 4

Waveguide properties

The propagation characteristics of the perfect fibre are discussed at length in the previous two Chapters. Here we introduce the fibre properties associated with practical fibres under operational situations. Fibre losses arise through material losses and radiative losses induced through mechanical imperfections. Bending of the fibre can introduce additional radiative losses. When fibres are joined together and when light is to be coupled to the fibre, fibre coupling losses and light launching efficiency are important and must be taken into account. At the same time the bandwidth of the perfect waveguide is usually modified by these perturbations.

4.1 Loss

A bound mode of a perfect fibre waveguide propagates with zero loss. In practice, the dielectric material is not lossless or perfectly homogeneous even if the fibre dimension is perfect. The bound mode will therefore be attenuated. The losses, however, are generally very small per unit wavelength of propagation distance, so that the propagation β derived by taking n_{co} to be real is valid. The actual loss can be simply computed by assuming the perfect waveguide field to be attenuated. In high-silica glasses the material losses can be grouped into two categories: absorption and scattering.

For silica based fibres, basic absorption is due to the electronic-absorption band edge of the silica host material at the ultra-violet-wavelength region and the molecular absorption edge of the silica host and its dopant at the infra-red region. In addition, impurity ions could contribute absorption bands. Impurity ions of transition elements such as Fe, Cu, Co and Mn have absorption bands in the near infra-red region. However, with concentration levels of the order of a few parts per billion, their contribution is lower than the scattering-loss component. The only impurity ion which contributes significant loss is OH^-. The hydroxyl ion at a concentration level of around 100 parts per billion contributes several dB/km loss at 1·39 μm wavelength. This is the second harmonic of the main absorption band at 2·73 μm and can only be reduced through a very careful drying process. A typical absorption spectra due to OH^- is shown in Fig. 4.1 for a pure quartz sample.

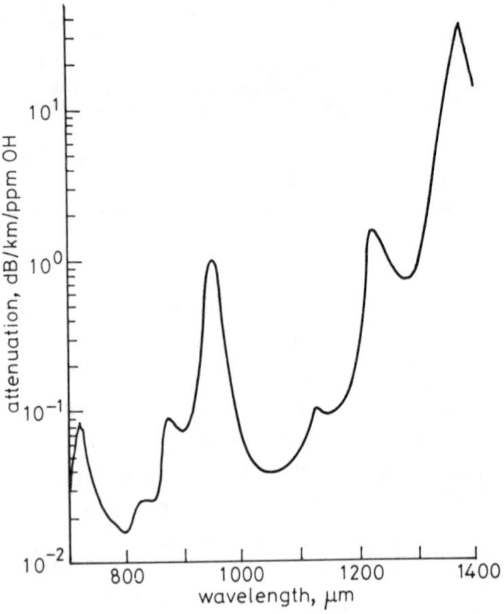

Fig. 4.1 *Typical absorption spectra due to OH^- for a pure quartz sample*

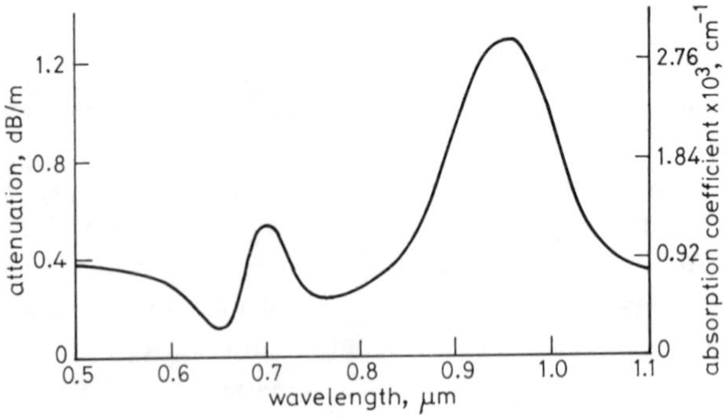

Fig. 4.2 *Early measurement of loss on bulk SiO_2 sample with low OH^- content* (Reproduced from Ref. 2)

An early measurement of loss on a bulk SiO_2 sample with low OH^- content is as shown in Fig. 4.2, which was, as it were, an existance proof that, by reducing impurity content, very low loss can be achieved. This SiO_2 sample was suspected to have negligible amount of transition elements, but had several parts in 10^6 of OH^-.

It was established later that the losses contributed by transition elements and OH^- vary with types of glass composition, the way the impurity ions are co-ordinated within the glass structure, and the state of oxidation.

Spectra losses measured for a glass are shown in Fig. 4.3 for the impurity levels tabulated. Variation of losses for different glasses with different impurity elements is shown in Table 4.1.

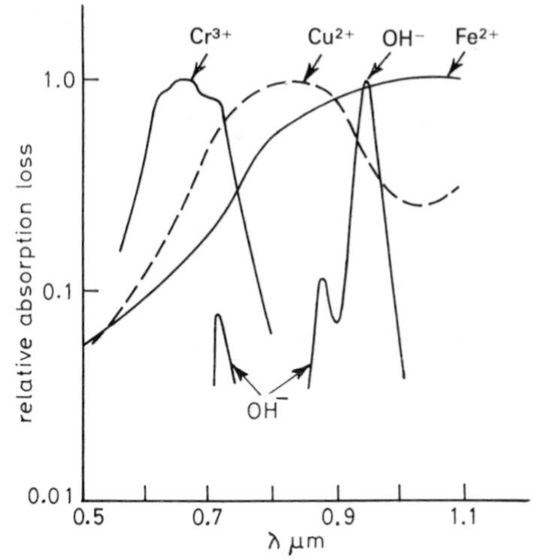

Fig. 4.3 *Relative absorption loss versus wavelength for certain ions in glass* (Reproduced from Ref. 3)

Table 4.1: *Variation of losses for different glasses with impurity elements* (Reproduced from Ref. 4)

Impurity element	Absorption (dB/km) induced at 850 nm for each part per million		
	Na_2O - CaO SiO_2*	Na_2O - B_2O_3 Tl_2O - SiO_2†	SiO_2 (fused silica)‡
Fe	125	5	130
Cu	600	500	22
Cr	10	25	1300
Co	10	10	24
Ni	260	200	27
Mn	40	11	60
K	—	40	2500

* NEWNS, G. R., PANTELIS, P., WILSON, L., UFFEN, R. W. J., and WORTHINGTON, R.: 'Absorption losses in glasses and glass fibre waveguides', *Opto-Electronics*, 1973, **5**, p. 289

† UCHIDA, T.: 'Preparation and properties of compound glass fibres', URSI General Assembly Commission VI, Lima, 18 August 1975

‡ SCHULTZ, P. C.: 'Optical absorption of the transition elements in vitreous silica', *Am. Ceramic Soc. J.*, 1974, **57**, p. 309

The ultra-violet absorption edge is due to electronic transitions. The absorption is expressible as

$$\alpha_{\mu\nu} \doteqdot \exp -(E - E_g)/\Delta E \qquad (4.1)$$

where E is the energy corresponding to the wavelength of the incident optical energy. E_g is the energy gap and ΔE is a constant for the material in question. The intrinsic ultra-violet absorption for SiO_2 with low GeO_2 doping has been estimated to be as shown in Fig 4.4. The infra-red absorption edge is due to molecular vibration bands. These are modified as dopants are added (see Fig. 4.5).

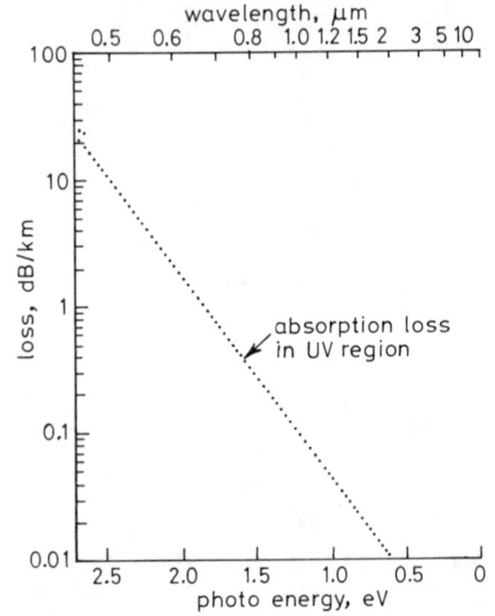

Fig. 4.4 *Spectral-loss curve of germania-doped silica-glass-core fibre inherent ultra-violet absorption loss*

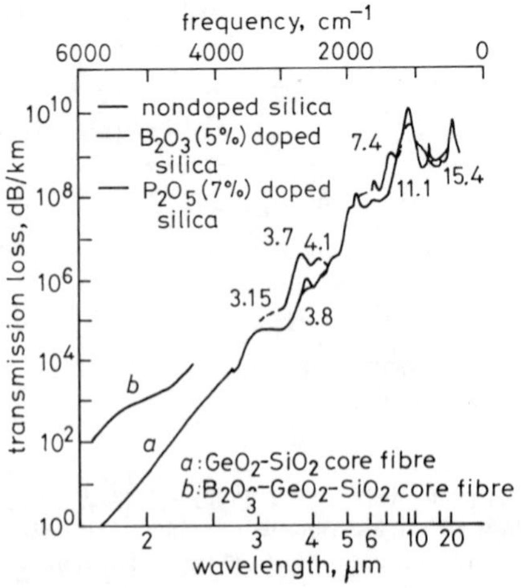

Fig. 4.5 *Absorption spectra for nondoped fused silica, doped silicas and fibres (Reproduced from Ref. 5)*

Scattering loss for fused SiO_2 or glass is due to microscopic inhomogeneities of material refractive index. As a glass is a super-cooled liquid, the micro-inhomogeneity is the frozen-in Brownian motion. The faster the glass is cooled the larger the scattering loss, while the addition of dopants would also increase the scattering loss.

For the scattering loss from a liquid, a quick calculation assuming only thermally induced fluctuation is given by

$$\alpha_s = \frac{8\pi^3}{3\lambda^4}(n^8 p^2)\, kT\beta \tag{4.2}$$

where

n = refractive index

p = photoelastic coefficient

k = Boltzmann's constant

T = absolute temperature

β = isothermal compressibility

For glass this expression is modified to

$$\alpha_s \propto \frac{T_F}{\lambda^4} f(n) \tag{4.3}$$

where T_F is the temperature at which glass is highly viscous, sometimes referred to as the fictive temperature, and $f(n)$ is a refractive-index dependence which is dependent on glass compositions.

For fused SiO_2, the scattering loss, corresponding to fibre quenching rate encountered during fibre drawing speed of about 1 m/s, is as shown in Fig. 4.6 for an experimental fibre, along with the instrinsic absorption bands. The spectral-loss increase due to an increase in dopant quantity is as shown in Fig. 4.7. From Fig. 4.6 minimum fibre loss occurs at about 1·5 μm. For a single-mode fibre with a fibre NA of about 0·06, minimum loss of 0·12 dB/km may be possible. The loss at 1·3 μm is then $0{\cdot}12 \times (1{\cdot}5/1{\cdot}3)^4$, which is around 0·22 dB/km.

If the λ^4 dependence of loss is maintained in other glass systems, the loss at longer wavelength due to scattering will be very much lower. The search is on for materials with a molecular absorption band well into the infra-red region.

It can be predicted from material considerations that many elements can form crystals and glass. Some elements have ultra-violet and infra-red absorption bands with minimum crossover loss at λ in the 2—10 μm region (see Goodman, Chapter 9.2). λ^4 dependence of scattering loss is a manifestation of the scattering centres being much smaller than λ. It is reasonable to expect that this law should hold for most glassy materials. Hence minimum loss can be as low as 0·001 dB/km around 4—5 μm if suitable material can be formed into a homogeneous glass. Chalcogenide and fluoride glasses have emerged as candidates, but despite purification of these glasses, their physical characteris-

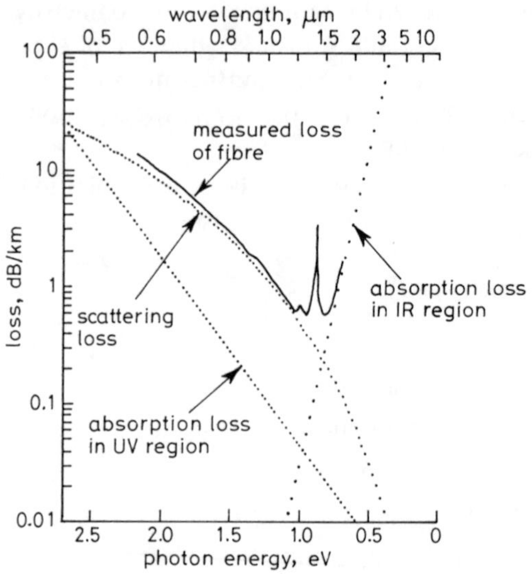

Fig. 4.6 *Spectral-loss curve of germania-doped silica-glass-core fibre separated into inherent ultra-violet absorption loss, scattering loss and inherent infra-red absorption loss* (Reproduced from Ref. 6)

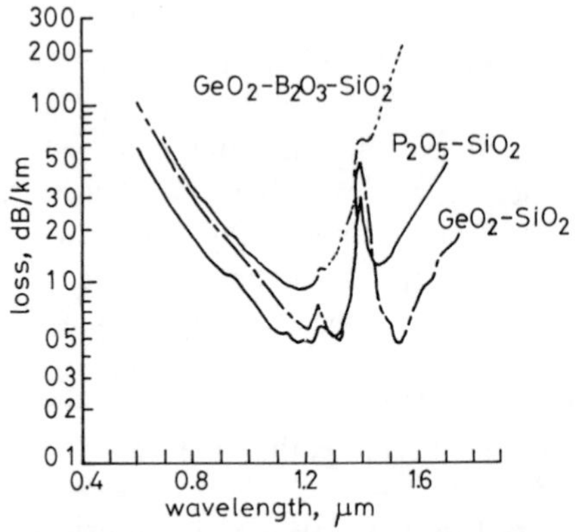

Fig. 4.7 *Spectral-loss curves of germania-doped borosilicate glass-core fibre, phosphosilicate-glass-core fibre and germania-doped-silica glass core fibre whose lengths are 1·2, 1·1 and 0·8 km, respectively* (Reproduced from Ref. 6)

tics are far inferior to SiO_2 glass. At this stage, the best loss reported is still higher than that of the SiO_2 system.

Spectra loss is practically constant over the operating temperature range encountered by fibres for all normal applications. It is however, affected by radiation from γ rays and neutron fluxes. The loss effect of radiation-induced loss is not predictable. The mechanisms involved are perhaps material stoichiometry, which is dependent or modified by the addition of dopants. The induced loss also can be bleached, in other words recovered, through optical and/or thermal annealing over time. Induced loss sometimes also shows saturation with dosage level. In general, the radiation effect is smaller at longer wavelengths. The studies carried out on silica fibres with different dopants showed that acceptable radiation-induced losses can be achieved for different applications. However, for very high radiation dosages of the order of 10^5 rads, the induced loss is unacceptably high over a significant time period of many minutes.

4.2 Fibre dispersion

To discuss fibre dispersion the fibres must first be classified into several practically important categories: large-core graded-index fibres, standard graded-index two-window fibres, single-mode fibre and dispersion compensated single-mode fibre.

For multimode fibre, fibre dispersion is dominated by intermodal delay; in the case of large-core graded-index fibre, a bandwidth of 100 MHz km or an RMS pulse broadening σ of 5 ns/km (taking bandwidth to be $1/2\alpha$). This type of fibre is principally designed for instrument-interconnect operation where connection lengths average well below 100 m. For minimising fibre/fibre joint loss without calling for extremely high mechanical accuracy of alignment, the fibre diameter is chosen to be conveniently large; 62·5 μm, 80 μm and 100 μm are some typical diameters. For good light-collection efficiency from light-emitting diodes with broad radiation lobe the NA is chosen to be as high as practicable; NA = 0·3 is a typical value. This type of fibre has a parabolic graded-index profile, and has a typical loss of about 4 dB/km at 0·85 μm and 2 dB/km at 1·3 μm. The standard-graded index fibre is designed for longer-distance interconnection applications. A standard fibre has a 50 μm core with a numerical aperture of 0·2. The core refractive index is graded with a q value of the index distribution, $n = n_{co}\,(1-2\Delta(r/\rho)^q)^{\frac{1}{2}}$, ρ chosen so that the optimum bandwidth occurs at a wavelength between 1·3 μm and 1·5 μm, so that the bandwidth at both these operational wavelengths is at least 800 MHz km. This type of fibre has a loss of around 0·5 dB/km at both 1·3 μm and 1·5 μm.

The bandwidth or dispersion of the graded-index fibre is influenced by several factors. First, according to ray theory, if the profile $n(r)$ can be adjusted, the meridian rays can be delay compensated. If skew rays are present, the optimised profile will not compensate for the propagation velocity of these rays. Differential delay in propagation will arise. Secondly, the influence of material dispersion can be significant when the fibre is operating with broad spectral sources, especially with LEDs which have a spectral width of greater than 600 Å. The magnitude of material dispersion is derived from refractive-index variation with wavelength (Fig. 4.8). Expressed in dispersion, the delay in ps/km nm is as shown in Fig. 4.9. A calculated dispersion versus q relationship is shown in Fig. 4.10.

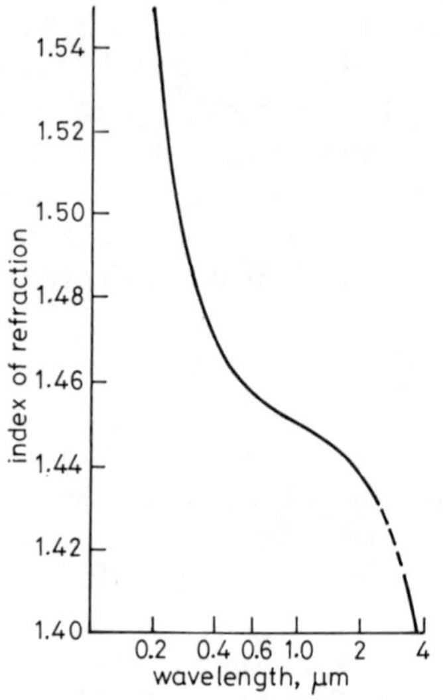

Fig. 4.8 *Variation in the index of refraction as a function of the optical wavelength for silica* (Reproduced from Ref. 7)

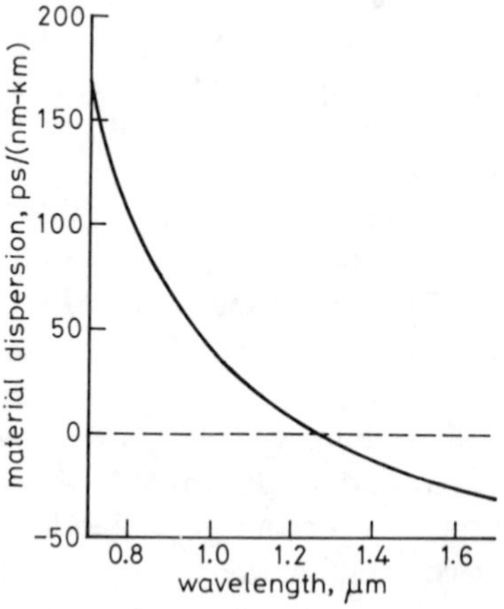

Fig. 4.9 *Material dispersion as a function of optical wavelength for silica* (Reproduced from Ref. 8)

Thirdly, the scattering and other imperfections within the fibre cause the propagating modes to exchange energy with each other and with the radiative mode. The mode-coupling effect, which will be discussed in more detail separately, results in an improvement in the average dispersion, since slow and fast modes are constantly being switched so that the slow modes become fast

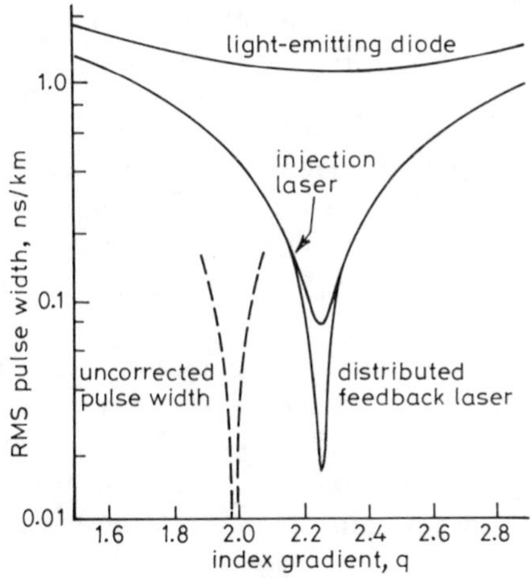

Fig. 4.10 *Calculated RMS-pulse spreading in a graded-index fibre versus the index parameter q at 900 nm*
The uncorrected pulse curve is for $\varepsilon = 1$ and assumes mode dispersion only. The other curves include material dispersion for an LED, an injection laser diode, and a distributed-feedback laser having spectral widths of 15, 1 and 0·2 nm, respectively (Reproduced from Ref. 9)

modes and vice versa. In fact, it can be shown statistically that, after a certain length of propagation with random coupling, the modes form an equilibrium set of modes and the dispersion then increases as $\sqrt{L}$ instead of L. It is to be noted that the equilibrium length for high-quality fibre is much longer than that for the poorer-quality fibre. The mode coupling effect is illustrated in Fig. 4.11.

It is clear that the optimisation of graded-index fibre is at best imperfect.

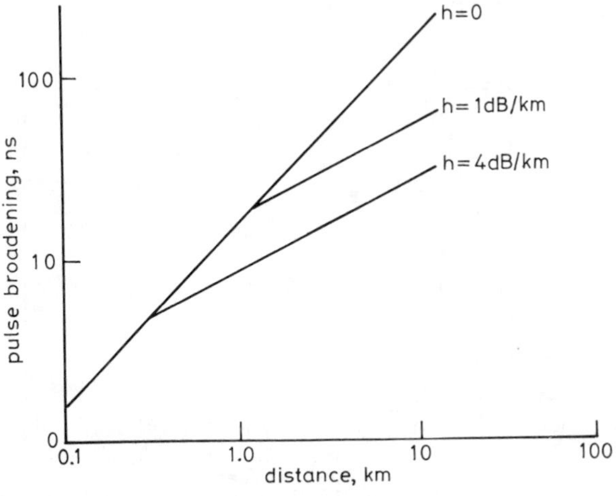

Fig. 4.11 *Mode-coupling effects on pulse distortion in long fibres for various coupling losses* (Reproduced from Ref. 10)

Indeed, the multimode fibre has one more serious problem. When operating with a source emitting light with significant coherence the modes can interfere with each other constructively and destructively. If discontinuities are present and as the signal modulates the light-source output, a signal-correlated noise will result from the multipath propagation, producing an interference pattern which is blocked by the imperfection. This demands the light source for multimode fibre to be less coherent. As a result, this fibre, which has been used extensively in transmission systems, is being superseded by the single-mode fibre.

A single-mode fibre is the principal general-purpose fibre and is an ideal transmission medium. A typical practical fibre has a core of 10 μm diameter and a numerical aperture of about 0·07 operating single mode at 1·3 μm. The fibre loss at 1·3 μm is about 0·4 db/km, and at 1·55 μm it is less than 0·2 dB/km. For a theoretical fibre with a uniform core refractive index the dispersion is defined as

$$\delta t = \frac{z}{2\pi C}\left\{\lambda^2 \frac{d^2\beta}{d\lambda^2} + 2\lambda\frac{d\beta}{d\lambda}\right\}\delta\lambda \tag{4.4}$$

with no material dispersion. This can be written in the form

$$\frac{dt}{d\lambda} = -\frac{z}{2\pi C} V^2 \frac{\delta^2\beta}{2V^2} \tag{4.5}$$

The dispersion for a single-mode waveguide arises due to the spread of the signal frequency. In other words, in an ideal single-mode fibre, the group velocity of the signal is fully characterised when the optical carrier is a truly single-frequency coherent wave. The dispersion can be expressed as a pulse spread in a convenient unit such as ps/km nm. The dispersion is a function of the operating V value of the waveguide (Fig. 4.12). The group velocity for the HE_{11}-mode versus V is shown in Fig. 4.13.

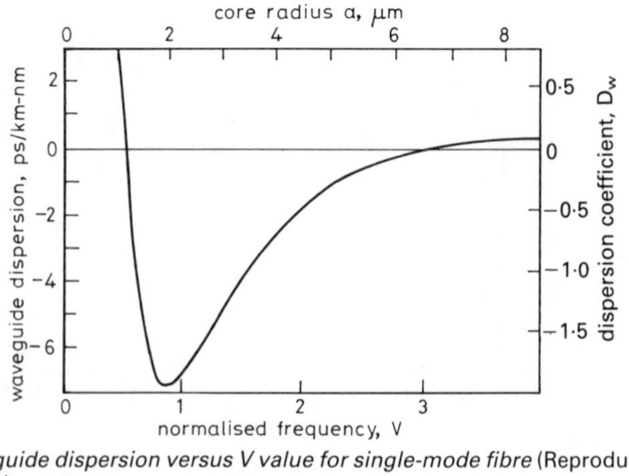

Fig. 4.12 *Waveguide dispersion versus V value for single-mode fibre* (Reproduced from Ref. 11)

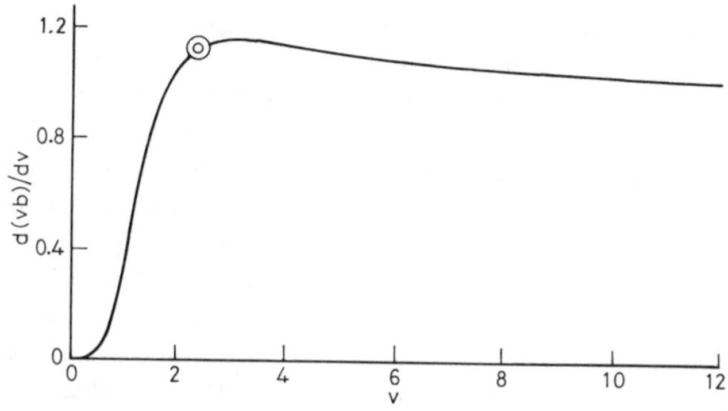

Fig. 4.13 *Normalised group delay d(vb)/dv as a function of v*

Signal bandwidth corresponding to 1 nm wavelength spread at 1 μm carrier wavelength is approximately 3×10^{11} Hz, corresponding to an RMS pulse width of ~2 ps. Fig. 4.12 shows that the single-mode fibre can handle a signal bandwidth over 1 km of 3×10^{11} Hz provided it operates near the cut-off point for the second mode of 2·4.

If the optical carrier has a significant line width, the waveguide dispersion is the contribution of the source line width and is not due to the signal bandwidth. On the other hand, the material dispersion shown earlier and replotted on the same scale (Fig. 4.14) shows that its contribution cannot be ignored. In fact, by pure coincidence the material system commonly used for single-mode waveguide, namely SiO_2 and SiO_2 with Ge doping, gives material dispersion in the wavelength band of interest of about equal magnitude but opposite in sign. Hence, a zero-dispersion region exists. The topic will be discussed shortly.

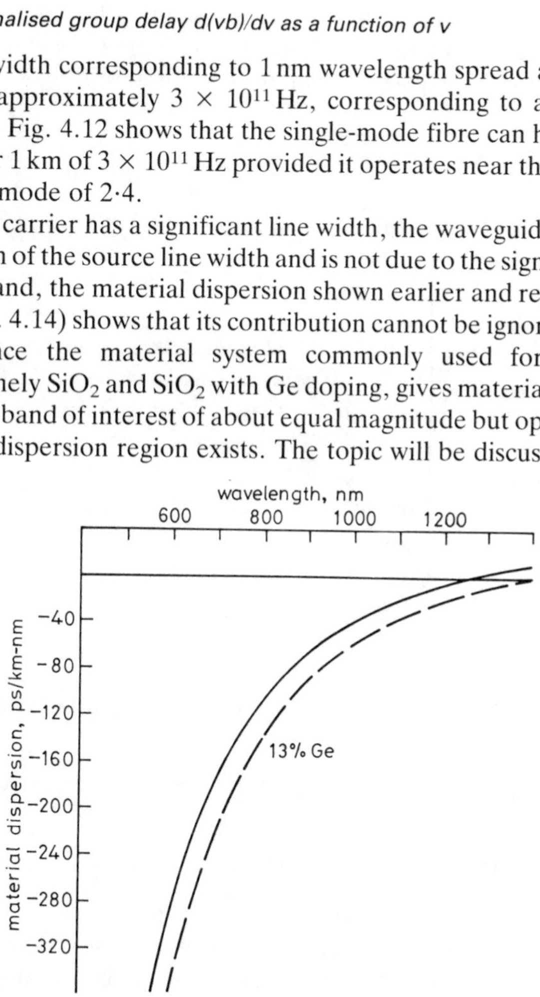

Fig. 4.14 *Material dispersion versus wavelength for Si- and Ge-doped Si* (Reproduced from Ref. 12)

In a practical single-mode fibre the refractive index is not uniform, as a result of the fabrication technique. Hence, the dispersion departs from that of the ideal waveguide. Several detailed studies have been undertaken for single-mode fibre with radially graded profile of refractive index. The legacy of multimode study provided the tools needed to achieve an early solution of a single-mode fibre with a parabolic index profile. This profile gives the first higher-order mode cut-off V to be at 3·5 instead of 2·4 (Fig. 4.15).

For shifting the zero-dispersion wavelength and maintaining low-loss propagation, the triangular profile was investigated. This design shifts the cut-off wavelength as well as changing the magnitude of the waveguide dispersion, so that, for the SiO_2/SiO_2 Ge system, the zero dispersion can be shifted to 1·55 μm, where the fibre loss is a minimum. For this type of fibre, low loss of 0·23 dB/km has been achieved. This compares reasonably well with a low-loss figure of <0·2 dB/km for an uncompensated fibre.

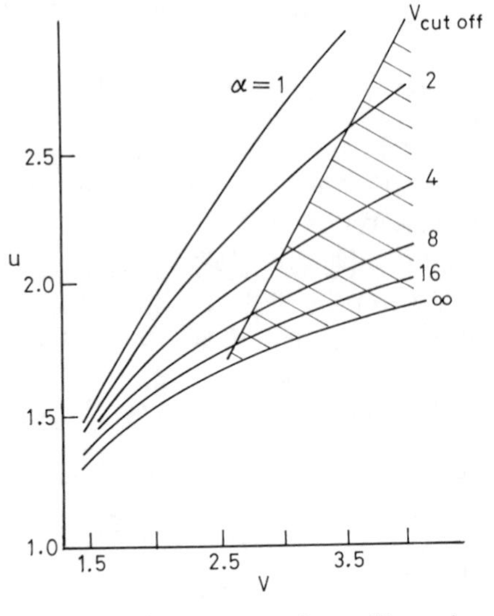

Fig. 4.15 *Cut-off loci for fibre with 'power-law' core* (Reproduced from Ref. 12)

Many practical fibres have a centre dip in the core refractive-index profile. By taking the index profile as

$$n(\rho) = n_{co}[1 - 2\delta(1 - \rho)^{\alpha}]^{\frac{1}{2}} \text{ where } \rho \text{ is normalised radius}$$

the modification of cut-off of V is as shown in Fig. 4.16.

In a practical fibre made by the MCVD process, the core index has a dip at the centre. This is caused by the evaporation of the Ge dopant during the formation of the fibre preform. The centre dip does not produce any significant effect on fibre performance. It can be interpreted that the centre dip

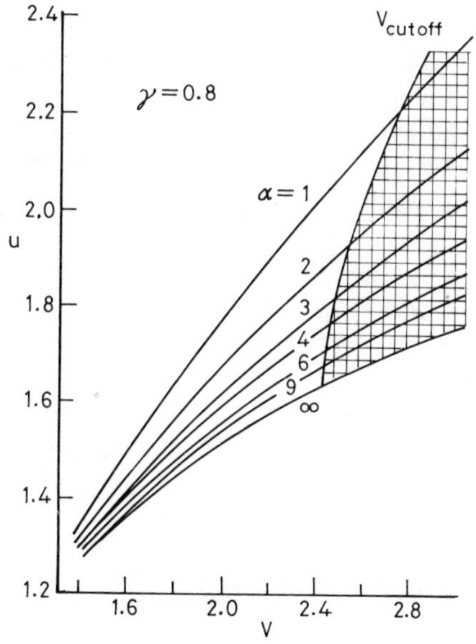

Fig. 4.16 *Radical eigenparameter u(v) behaviour for fibre with inhomogeneous core having a central dip* (Reproduced from Ref. 12)

will be seen by the wave as a decrease in effective index of a uniform core, and thereby results in a change in the effective diameter of the fibre core.

Another variation is the *W*-fibre or depressed-cladding fibre (Fig. 4.17). The influence of the depressed cladding is to modify the cut-off wavelength and the shape of the group-velocity v/s *V* curve. The relevant relationships are shown in Figs. 4.18—4.20. Hence it is possible to design the waveguide to tailor a near-zero dispersion region.

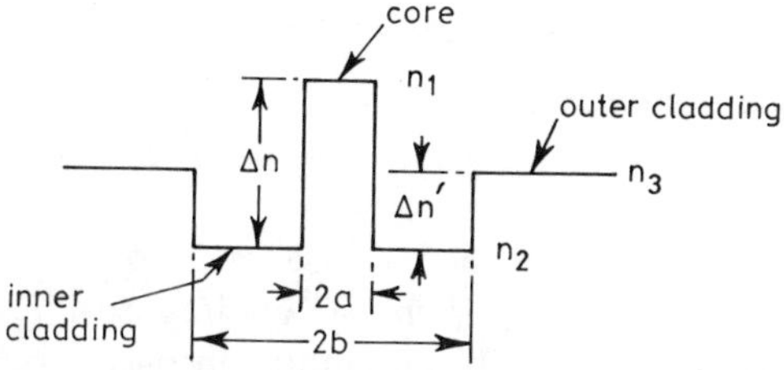

Fig. 4.17 *Refractive-index profile schematic for a W-fibre, showing inner cladding with depressed index*
index contrast R = $\Delta n'/\Delta n$
diameter ratio c = b/a
(Reproduced from Ref. 12)

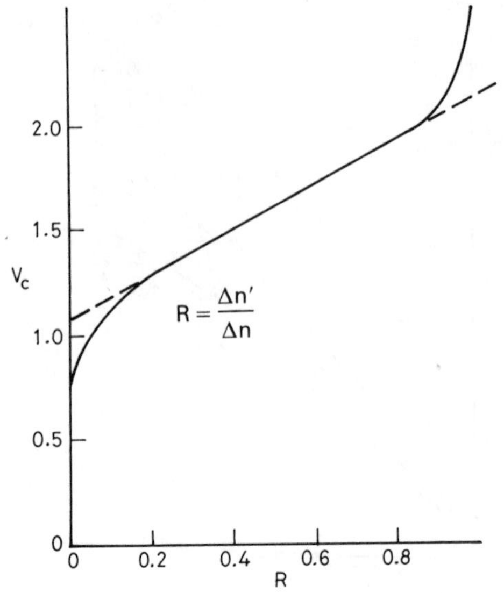

Fig. 4.18 *Dependence of the (non-zero) LP01 cut off in a W-fibre upon the index, contrast ratio R for very large inner cladding width* (Reproduced from Ref. 12)

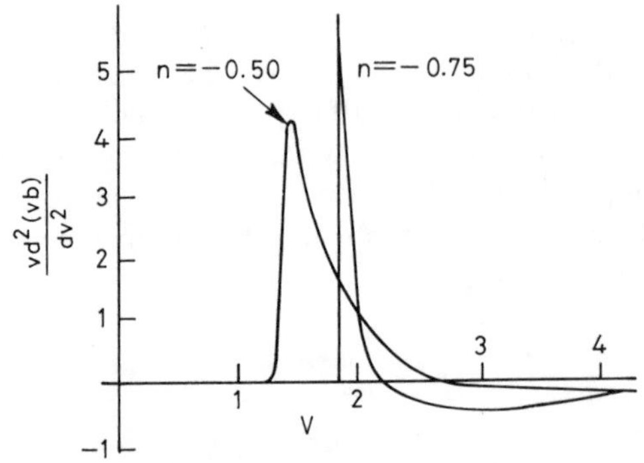

Fig. 4.19 *Principal contribution $vd^2(vb)/dv^2$ for a W-fibre* (Reproduced from Ref. 12)

This can be generalised further to a segmented core-fibre. (Figs. 4.21 and 4.22). All these modified fibres, however, impose three operational conditions. One is the increase in fibre-manufacturing tolerance requirement and the second is an increase in transmission loss due to a small increase in scattering loss. The third is a more stringent splicing- or jointing-tolerance requirement. Since the information-bandwidth–distance product is of the order of 10^{11} Hz km, an uncompensated fibre operating with a narrow-line-

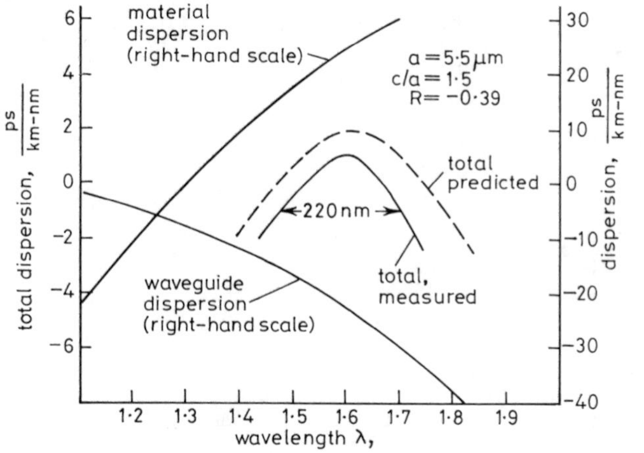

Fig. 4.20 *Material, waveguide, and total dispersion in an experimental W-fibre* Bandwidth 220 nm for ±1 ps/km-nm (Reproduced from Ref. 12)

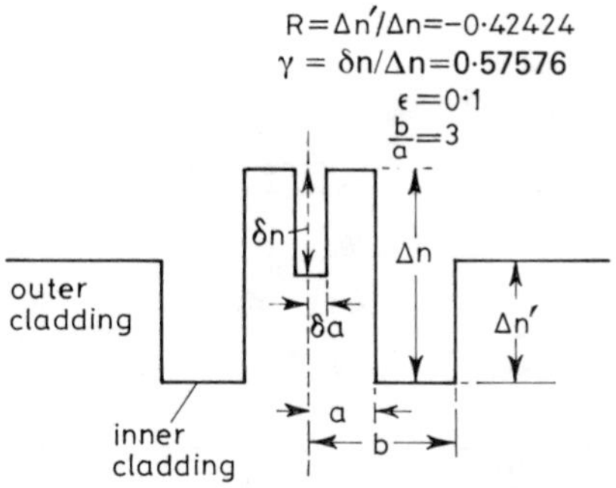

Fig. 4.21 *Cross-section of step-type W-fibre with rectangular dip, showing the defined dimensionless parameters* (Reproduced from Ref. 12)

width optical source offers a good solution for most long-distance trunking and intercity transmission applications. For special applications where the longest repeater spacings are mandatory, such as repeaterless submarine systems and trans-continental links, the special fibre optimised for maximum bandwidth–distance product could be justified.

A repeater-spacing–bandwidth/system-bit-rate graph (Fig. 4.23) is effective to illustrate the regions appropriate for different fibre types and application areas.

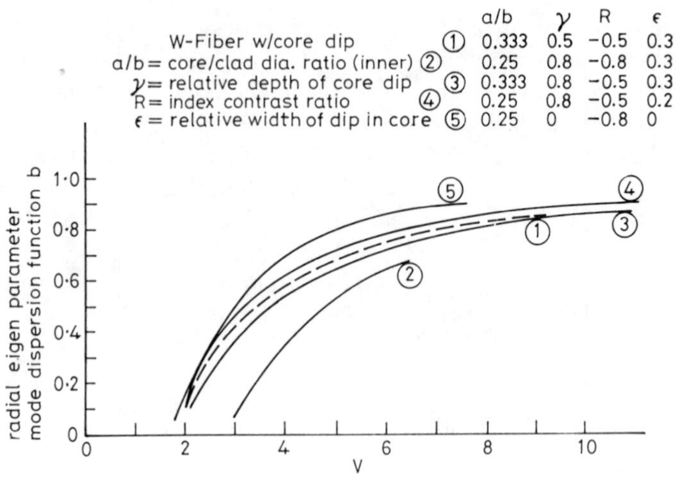

Fig. 4.22 *Fundamental mode dispersion function b for W-fibres showing effect of rectangular central dips*
b = o is cut-off
b is a normalised group velocity
(Reproduced from Ref. 12)

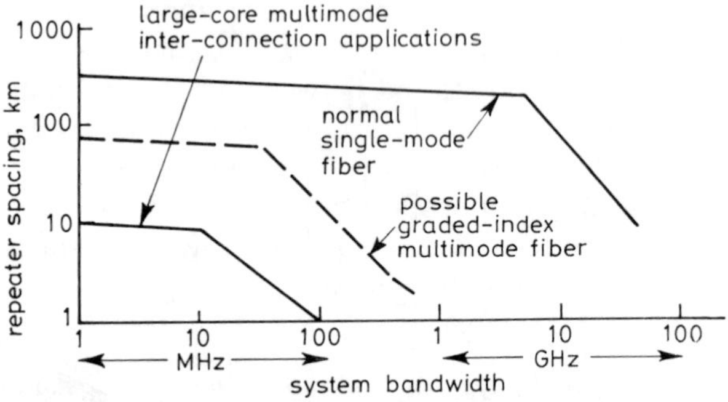

Fig. 4.23 *Repeater-spacing v/s bandwidth/system bit-rate*

One other way for shifting dispersion compensation is to select or modify the material refractive-index/frequency characteristics and the operating V value. For silica glass plus dopants $n(\lambda)$ is expressible as

$$n^2 = 1 + \sum_{j=1}^{3} A_j \, \lambda^2/(\lambda^2 - \lambda_j^2)$$

where A_j and λ_j are specific for each glass and are measured values. The three sets of constants are called Sellmeier constants. By choosing the fibre diameter, the V value can be adjusted.

From the foregoing discussion it is evident that single-mode-fibre design for particular characteristics can be attempted. The design criteria are for low loss and dispersion. The parameters at our disposal are material composition, NA, core diameter, core and cladding refractive indices, and core-index profiles. It can be envisaged that the parameters are not mutually independent, and will also affect the fibre loss. Hence, a compromise is called for when the highest bandwidth, or lowest loss, is to be achieved. Simultaneous achievement of both is not necessarily possible, particularly if the operating wavelength is fixed. Design by iteration is the usual approach. Attempts at solving the reverse problem of being given a bandwidth requirement to solve for the core refractive-index profile have been made with limited success.

4.3 Waveguide imperfections

Scattering and absorption centres along the fibre give rise to loss and mode coupling. Besides these intrinsic imperfections, there are several classes of extrinsic imperfections: scattering at the core/cladding interface, scattering due to bubble or particle inclusion within the fibre, and cross-sectional dimensional changes. These are z-direction-dependent imperfections. Longitudinal dimensional variation and abrupt cross-sectional waveguide parameter changes are other z-direction-dependent discontinuities. The effects of these imperfections are to be examined in some detail as a couple of different classes of problems with different solutions. In general, the imperfections cause mode coupling between guided modes and between guided and radiative modes, both in the forward and backward directions. Hence they result in changes of loss and dispersion characteristics. Also, in general, it is difficult to obtain rigorous electromagnetic solutions for these problems.

The first class of imperfections are randomly distributed scattering centres. These could be due to core/cladding-interface imperfections, in which case the scattering takes place at the core/cladding boundary. They could also be due to particle or bubble inclusion. Using the ray optics approach for multimode waveguide, the scattering centres lead to power redistribution in bounded, leaky and radiative rays. Assuming that scatterers result in forward scattering (this is true if centres are small), their distributions are random, and the ray directions at the beginning and end of the fibre are not correlated, then all rays are likely to scatter in all directions. With many rays the transit time is the RMS average of the transit times of these rays. Statistically this will result in the transit-time spread being proportional to $\sqrt{z}$.

If the scattered energy of a ray is mainly confined to a small cone angle in the forward direction, the ray diffuses outwards and generates rays with angles bordering on its direction of propagation. In modal parlance the mode couples to its near neighbours only. It can be shown that $I(\theta_\phi, \theta_z, z)$, total power per unit solid angle of all rays with direction angle θ_ϕ, θ_z over the whole cross-section at positive z, satisfies a diffusion equation.

Formally, isolated discontinuities are studied as a current dipole within a waveguide. Each dipole generates an E and H field expressible in terms of bounded modes and radiation field propagating in the forward direction on one side of the current source, and the backward direction on the other side. Only a few exact solutions can be constructed for either a transverse or a longitudinal dipole within a specific waveguide. To treat practical discontinuities or imperfections, a tubular current-source concept has been used (Fig. 4.24). The current vector is given as

$$J = \hat{n}\, g(z) \frac{\delta(r - r_0)}{2\pi r_0} \cos l(\phi - \phi_0) \text{ for } -L \leq z \leq L \tag{4.7}$$

The case $g(z) = e^{icz}\sin\Omega_z$ describes a periodic imperfection. It can be shown that, for such a case, the bounded-mode power generated is small except at near resonance defined by $\Omega = |C \pm \beta_l|$; i.e. when the discontinuity phase constant and phase constant of the lth mode resonates with the spatial periodicity of the discontinuity. The radiation field generated is also small except near resonance at a particular direction of radiation θ given by $|C \pm kn_{cl}\cos\theta| = \Omega$. The power in the bounded mode is proportional to the square of the tube length, and the power in the radiation field is linearly proportional.

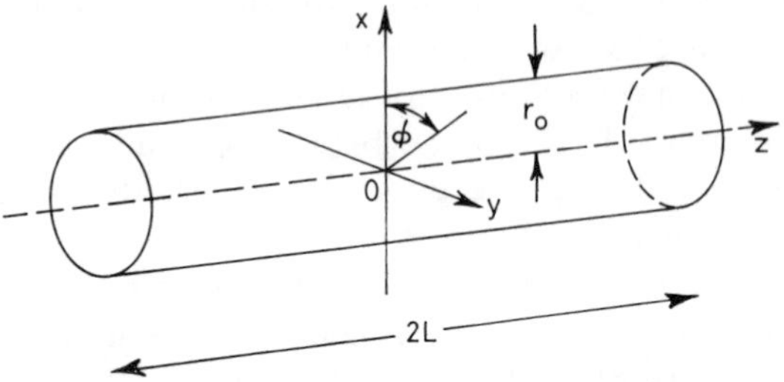

Fig. 4.24 *Tubular current source of length 2L and radius r_0 carries the distribution of J of eqn. 4.7 on its surface* (Reproduced from Ref. 1 of Chapter 2)

The physical interpretation of general abrupt discontinuities is therefore as follows. The EM wave incident at the discontinuity causes the discontinuity to appear as a current source which generates forward and backward bounded modes and radiation fields. An abrupt discontinuity can be envisaged as a delta function, and can be represented as an infinite set of spatial Fourier components. Hence, the discrete discontinuity can be treated as a sum of sinusoidal discontinuities spatially distributed throughout the fibre length. Extrapolating from the result of the study of a single sinusoidal distribution of imperfection, a discrete discontinuity is expected to generate all modes. From geometrical ray-optics considerations a discontinuity scatters a ray in many directions. If many omnidirectional scatterers exist, or even if scattering is into a small solid angle at each scatterer, the rays will constantly exchange energy with each other, so that the RMS average velocity spread will be proportional to $\sqrt{z}$. This random discontinuity will lead to power loss and modify dispersion, so that dispersion is proportional to $\sqrt{z}$ in the limit.

The second class of discontinuities is continuously varying imperfections in the z direction. This embraces a large number of types. The periodic dimension variations along z, or refractive-index variations along z, is a practical situation. This was amenable to theoretical studies, as indicated before, through the use of a current-tube model or through a coupled-mode formulation. Initial attempts were based on obtaining a solution of the waveguide with a sinusoidally varying boundary. A slow taper is another practical situation which can be dealt with theoretically. These two types of continuously varying imperfections can be used to describe any continuously varying discontinuities.

The sinusoidal variation of fibre dimension or refractive index gives rise to strong mode coupling between two modes such that Ω, the mechanical period, is equal to the difference of the phase constants, $|\beta_a - \beta_b| = \Omega$. In a multimode waveguide this difference cannot exceed $kn_{co} - kn_{cl}$; hence $\Omega \leqslant V/\rho\sqrt{\Delta/2}$, can only cause significant radiation $2{\cdot}4 \times 10^4$ cycles per metre, or a period of $0{\cdot}5$ mm for $\rho = 5\,\mu$m, $\Delta = 0{\cdot}005$ and $V = 2{\cdot}4$.

The coupling between modes instigated by dimensional variations is therefore caused by the magnitude of the frequency component of the Fourier spectrum of the mechanical deformation corresponding to Ω. These are generally the higher-order spatial harmonics of the actual mechanical imperfections, which usually have physical dimensions of many millimetres. The coupling to radiation field is governed by $|\beta_a - kn_{cl}\cos\theta| = \Omega$, for $0 < \theta < \pi$, i.e. $\beta_a - kn_{cl} < \Omega < \beta_a + kn_{cl}$ for a single-mode fibre as above but with $\rho = 5\,\mu$m, $2\pi/\Omega = 0{\cdot}5$ mm, ripple period or greater will cause negligible radiation loss for the fundamental mode operating at $V = 2{\cdot}4$.

4.4 Coupled mode theory

The coupled-mode theory provides a physical description of the modal propagation in fibres. Except for systems with simple coupling, the solution is difficult to construct analytically. For treating simple fibre-imperfection problems the coupled-mode theory has been applied with some success. The relationship between the amplitude of the modal components is

$$\frac{da_{+j}}{dz} - i\beta_j a_{+j} = \sum_k (K_{jk}^{++} a_{+k} + K_{jk}^{+-} a_{-k}) \tag{4.8}$$

$$\frac{da_{-j}}{dz} + i\beta_j a_{-j} = \sum_k (K_{jk}^{-+} a_{+k} + K_{jk}^{--} a_{-k}) \tag{4.9}$$

with K_{jk}, the coupling coefficient, which is the integral across an infinite aperture of the scalar products of the tangential and longitudinal E and H vectors of the modes concerned.

These equations are exact equivalents of Maxwell's equation. If coupling to backward modes is neglected and only one mode is assumed to exist at $z = 0$, the coefficient for kth mode is given approximately by

$$a_k^+(z) = a_j^+(0)e^{-j\beta_k z} \int_0^z K_{jk}^{++}(z')e^{j(\beta_k - \beta_j)z'}dz' \tag{4.10}$$

Hence coupling is governed by the Fourier component of K_{jk} at the spatial frequency $(\beta_j - \beta_k)$ only for the jth and kth mode. From this it can be deduced that a strictly periodic coupling mechanism will couple two modes only with a beat frequency given by $2\pi/(\beta_j - \beta_k)$.

By considering ensemble average power for each mode, $P_j = <|a_j|^2>$, the coupled-mode equations can be simplified to a finite system of coupled-power equations. The coupling problem of a fibre with a random set of coupling points, such as those introduced by bends, has been solved. The pulse width is shown to increase with $\sqrt{z}$. Radiation loss caused by random bends depends on the spectrum of the fibre-axis distortion function. Severe radiation loss can occur with distortions containing significant Fourier components equal to the beat length of the propagating mode, and is in a specific radiation direction. Comparison of loss for single-mode fibre with multimode fibre for the same distortion functions shows that the single-mode fibre will radiate more strongly. However, the difference in fibre geometry makes practical single-mode fibre less sensitive to microbends. This aspect will be discussed.

4.5 Waveguide bend

Waveguide bend is a difficult EM-theory problem. It can be argued that, for a planar waveguide carrying a bounded mode, as the waveguide is bent into a uniform curve, the phase front cannot be plane across an infinite cross-sectional plane (Fig. 4.25). Somewhere away from the centre of the radius of curvature, the phase front must travel at higher than the velocity of light in that medium. This means that no bounded mode exists in a curved guide.

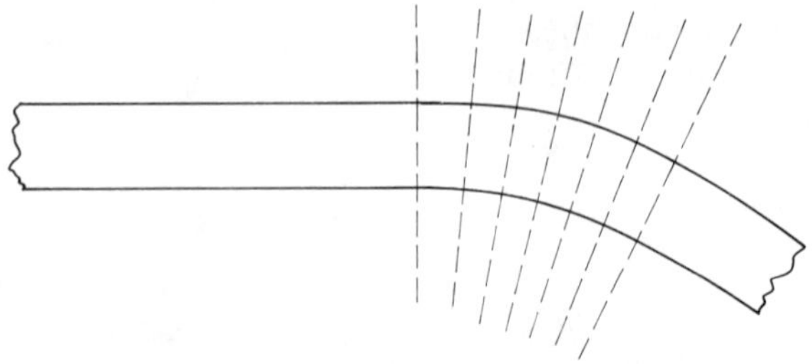

Fig. 4.25 *Phase front for waveguide but into uniform curve*

A ray approach can be used to describe the bent planar waveguide. All rays are leaking. The radiation caustic r_{rad} is given, from eqn. 3.3 with $\bar{\beta} = 0, \bar{l} = \bar{l}_b$ $\rho = R + \rho$, by

$$r_{rad} = (R + \rho)\bar{l}_b/n_{cl} \tag{4.11}$$

where $R_b = \{r/(R + \rho)\}n(r)\cos_\phi(r)$ from eqn. 3.29 (see Fig. 4.26).

For a planar waveguide with step profile

$$\bar{l}_b = n_{co}\cos\theta_\phi = \{(R - \rho)/(R + \rho)\}\, n_{co}\cos\theta_{\phi} \tag{4.12}$$

where θ_ϕ and $\theta_{\phi'}$ are as shown in Fig. 4.27

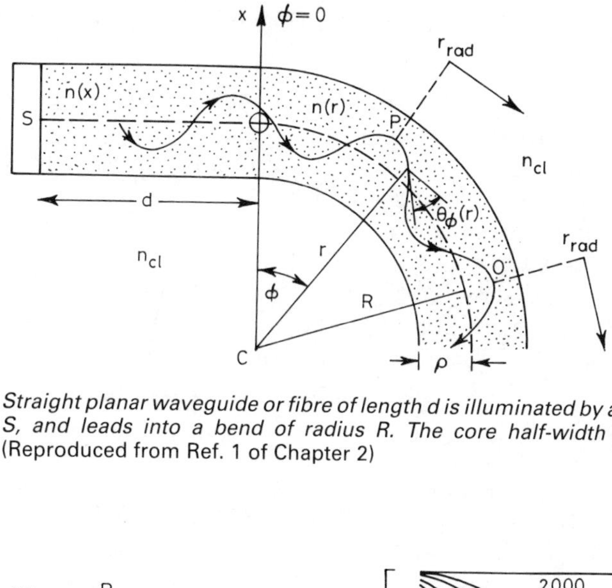

Fig. 4.26 *Straight planar waveguide or fibre of length d is illuminated by a diffuse source S, and leads into a bend of radius R. The core half-width or radius is ρ* (Reproduced from Ref. 1 of Chapter 2)

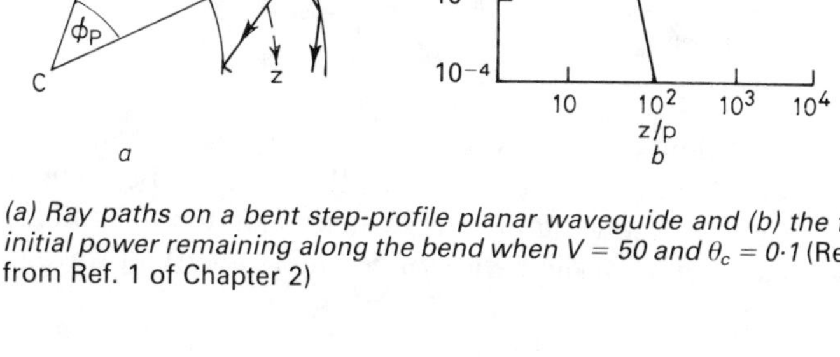

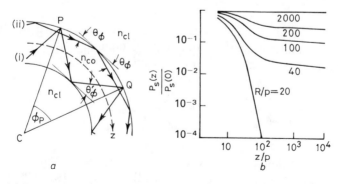

Fig. 4.27 *(a) Ray paths on a bent step-profile planar waveguide and (b) the fraction of initial power remaining along the bend when V = 50 and θ_c = 0·1* (Reproduced from Ref. 1 of Chapter 2)

Hence from eqn. 4.11

$$\frac{r_{rad}}{R+\rho} = \left(\frac{n_{co}}{n_{cl}}\right)\cos\theta_\phi \tag{4.13}$$

Thus, if $\theta_\phi < \theta_c$, it tunnels, while, if $\theta_\phi > \theta_c$, the ray refracts. Rays lose power only at the outer interface. Calculating the transmission coefficient for tunnelling rays, we use numerical computation:

$$T = \exp\left\{-2k\int_{r_{tp}}^{r_{rad}} [\beta^{-2} + \bar{l}^{-2}\,\frac{\rho^2}{r^2} - n^2(r)]^{\frac{1}{2}}\right\}$$

The power in the guide is seen to be significantly affected only until $R/\rho < 10^3$ for both a step (Fig. 4.27) and a graded (Fig. 4.28) bent planar waveguide.

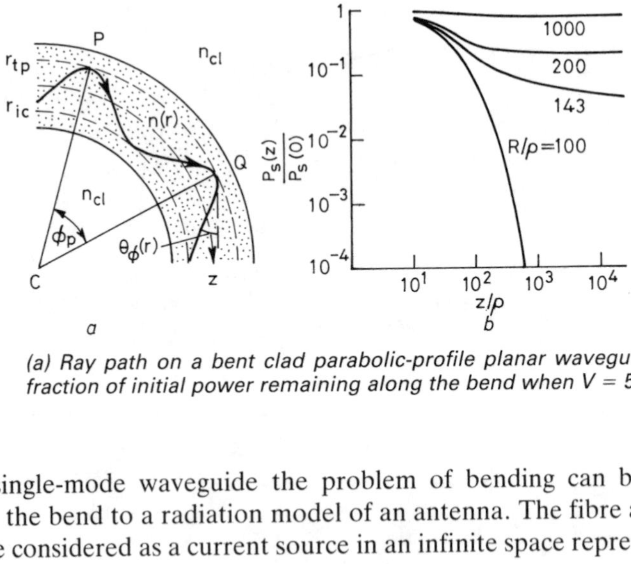

Fig. 4.28 *(a) Ray path on a bent clad parabolic-profile planar waveguide and (b) the fraction of initial power remaining along the bend when V = 50 and θ_c = 0·1*

For single-mode waveguide the problem of bending can be tackled by relating the bend to a radiation model of an antenna. The fibre and its modal field are considered as a current source in an infinite space represented by the cladding region. The current distribution is

$$\boldsymbol{J} = i\left(\frac{\varepsilon_0}{\mu_0}\right)^{\frac{1}{2}} k\,(n_{cl}^2 - n_{co}^2)\,\boldsymbol{E} \tag{4.14}$$

with E the exact field of the waveguide. Assuming large bend curvature such that the waveguide mode is not affected by the bend, the bending loss has been calculated, and is as shown in Fig. 4.29. In addition to the bend loss there is a transition loss when a straight waveguide enters a curved waveguide or any abrupt change of curvature. The transition loss between the straight and curved section is

$$\left(\frac{\rho}{R_c}\right)^2 \left(\frac{r_0}{\rho}\right)^6 \frac{V^4}{8\Delta^2} \tag{4.15}$$

and between two curvatures the loss is proportional to

$$\left(\frac{R_1 \pm R_2}{R_1 R_2}\right)^2 \frac{\rho^2 V^4}{8\Delta^2} \left(\frac{r_0}{\rho}\right)^6 \tag{4.16}$$

Note that the transitional losses are significantly larger than the bend losses, which in practice are kept at a negligible level through the avoidance of small bend radii.

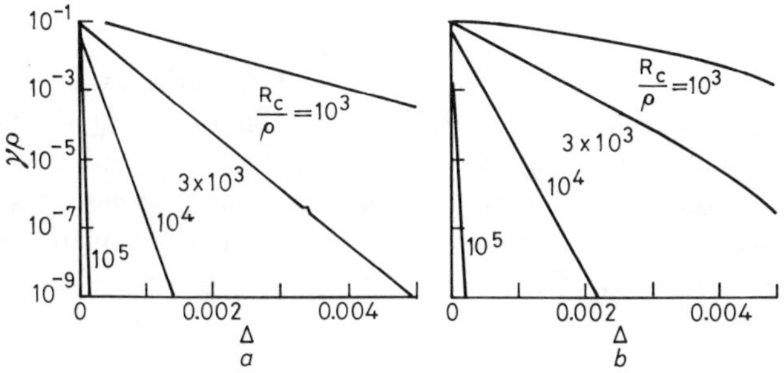

Fig. 4.29 *(a) Normalised attenuation coefficient γρ of eqn. 4.14 as a function of Δ for various values of the normalised bending radius R_c/ρ when the fibre has a step profile. (b) The corresponding results for a Gaussian-profile fibre* (Reproduced from Ref. 1 of Chapter 2)

4.6 Taper

Taper belongs to the general class of fibre imperfections which is slowly varying along z. Slowly varying is defined as changes which can be neglected over a distance z_p which permits a ray to undergo two reflections or a half period. Under this situation the taper can be treated as a series of uniform fibre sections within each of which no z variation occurs. In a multimode waveguide the taper can be treated and understood through ray tracing.

The linear taper with collimated incident beam illustrates the property of the taper as a section to convert ray angles and to concentrate ray power over a smaller area (Fig 4.30/*a*).

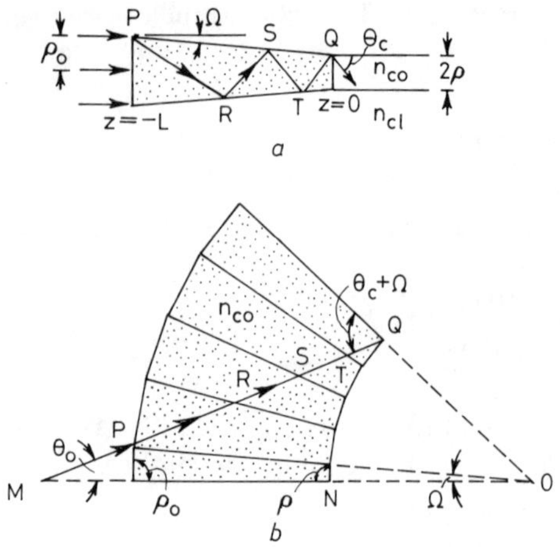

Fig. 4.30 *(a) Step-profile linear taper with taper angle Ω and (b) the equivalent geometrical path* (Reproduced from Ref. 1 of Chapter 2)

For a single-mode fibre taper the slowly varying criteria allow the field at each intersection of two waveguides to be matched by expanding the field in each, to the corresponding modes of the uniform sections. These modes are sometimes referred to as local modes. The field matching assumes that a local mode power is conserved. The slow variation criteria must ensure that length is adequate for beat length effect to be avoided. This is generally the case for all practical tapers.

4.7 End-face launching

Light is launched into the fibre by illuminating the end face of a fibre with light from a light source. Formally this is a field-matching problem. The incident field is known. It is to be matched to the modes excited in the fibre and the reflected field. For a rigorous solution the reflected field is part of the unknown to be solved. This makes the problem untractable. For a weakly guided fibre the reflected field can be assumed to be the same as that generated when the incident fields meet a uniform dielectric boundary. With this assumption the incident field in matched to the forward propagating components of the fibre field by considering either the tangential $\boldsymbol{E}$ or $\boldsymbol{H}$ field. Thus,

$$\boldsymbol{E}_t(x,y) = \sum_{jC} a_j\, \boldsymbol{e}_{t_j}(x,y) + \boldsymbol{e}_{t_r}(x,y) \tag{4.17}$$

$$\boldsymbol{H}_t(x,y) = \sum_{j} a_j\, \boldsymbol{h}_{t_j}(x,y) + \boldsymbol{h}_{t_r}(x,y) \tag{4.18}$$

Since the modes are orthogonal to each other, the mode coefficients can be obtained by taking cross products $\boldsymbol{h}_{t_k}{}^*$ or $\boldsymbol{e}_{t_k}{}^*$ and integrating over the infinite apertures.

$$a_k = \frac{1}{2N_k}\int_A \boldsymbol{E}_t \times \boldsymbol{h}_{t_k}{}^* \cdot \hat{\boldsymbol{z}}\, dA$$

or

$$= \frac{1}{2N_k}\int_A \boldsymbol{e}_{t_k}{}^* \times \boldsymbol{H}_t \cdot \hat{\boldsymbol{z}}\, dA \tag{4.19}$$

The power P_k is given by

$$P_k = |a_k|^2 N_k$$

The total power of the guided or bounded modes is

$$P = \sum_k P_k = \sum_k |a_k|^2 N_k \text{ with } N_k = \tfrac{1}{2}\left|\int_{A\infty} \boldsymbol{e}_k \times \boldsymbol{h}_k{}^* \cdot \hat{\boldsymbol{z}}\, dA\right| \tag{4.20}$$

N_k is a normalisation factor (see Table 2.5).

$\boldsymbol{E}_t$ or $\boldsymbol{H}_t$ is the same as those of the incident wave minus the Fresnel reflection term, since the dielectric boundary is assumed to be between n_i, which is usually air, and n_{co}, the core index, which is taken to be equal to n_{cl} at $z = 0$.

If the incident angle is normal to the end face the transmitted angle is still along the fibre axis, and the transmission power coefficient

$$T = 4n_i n_{co}/(n_i + n_{co})^2 \tag{4.21}$$

This expression holds approximately also for small angles of incidence $\theta_z = \theta_i \ll 1$. The incident power in this case can be taken as fully transmitted.

The field of the illuminating beam is usually a plane wavefront with a Gaussian power distribution, with

$$\boldsymbol{E}_i = E_x \hat{\boldsymbol{x}} = \hat{\boldsymbol{x}}\, f(r) \exp\left[ikn_i(x\theta_i + z)\right]$$

$$\boldsymbol{H}_i = (\varepsilon_0/\mu_0)^{\frac{1}{2}}\, n_i\, \hat{\boldsymbol{z}} \times \boldsymbol{E}_i$$

$$f(r) = \exp\left(-\frac{1}{2}\frac{r^2}{\rho_s{}^2}\right) \text{ and } P_i = \pi n_i\left(\frac{\varepsilon_0}{\mu_0}\right)^{\frac{1}{2}}\int r f^2(r)dr = \frac{\pi}{2}\left(\frac{\varepsilon_0}{\mu_0}\right)^{\frac{1}{2}} n_i \rho_s{}^2$$

where ρ_s is the spot radius.

The circular symmetry of the fibre allows the polarisation of the modes excited to be dictated by the polarisation of the incident beam. Matching the incident field to the modal field

$$P_l = \frac{\rho n_{co}}{4\pi}\left(\frac{\varepsilon_0}{\mu_0}\right)^{\frac{1}{2}} \left|\int_0^{\infty}\int_0^{2\pi} E \times \psi_l\, R dR d\phi\right|^2 \Big/ \int_0^{\infty} F_l^2(R)\, R\, dR \tag{4.22}$$

where

$$\psi_l = F_l(R)\cos l\phi \text{ or } F_l(R)\sin l\phi$$

at normal incidence $\boldsymbol{E}$ is independent of ϕ and $E_x = \exp(-r^2/2\rho_s^2)$
Hence only $l = 0$ in HE_{lm} modes are excited. For a graded-index core or single-mode fibre,

$$\psi_0 = F_0 = \exp(-r^2/2r_0^2) \text{ with } r_0 = \rho/V^{\frac{1}{2}}$$

the mode spot size

Hence

$$\frac{P_0}{P_i} = \left\{\frac{2\rho_s r_o}{\rho_s^2 + r_o^2}\right\}^2 \tag{4.23}$$

is the fraction of power launched in the fundamental HE_{11} mode. If $r_0 = \rho_s$ then all power is launched in the fundamental mode. Launching efficiency for each mode is therefore dependent on the closeness of fit of the incident tangential field and the modal field. The discussion applies to a single-mode waveguide since the fundamental modal field can be approximately described by a Gaussian beam. Hence the expression P_0/P_i is the launching efficiency.

If the incident beam is tilted the incident field has a ϕ dependence, and hence $l \neq 0$ modes will be excited. This is also the case if the beam is offset from the axis. It also can be shown that launching efficiency decreases exponentially with the offset. The effect of the presence of a lens is to change ρ_s if the convergence is not excessive. Otherwise E_x will be on a curved phase front, resulting in lower launching efficiency than expected from the plane-wave-front assumption. The description applies equally to the case when a fibre is butt jointed to another. The launching efficiency due to mismatch in size, tilt and offset can be computed as in the case of launching efficiency from an incident beam. The expression for mismatch tilt and offset are given in Table 4.2 for single-mode waveguides.

For multimode waveguide a geometrical-optics approach is more appropriate. A source is then represented as follows. The source of area dA emits within a specific core angle with intensity defined per unit solid angle and unit area. Hence

$$dP = I(\theta)d\theta dA$$

A diffused (Lambertian source) emits with $0 \leqslant \theta \leqslant \pi/2$. An LED can be a diffused source if it is not a super-radiant edge-emitting type. Using the geometrical-optics approach the launching problem is treated as the coupling of the incident power that arrives at the end face of the fibre which lies within the acceptance angle of the fibre. It can be readily shown that launching from a diffuse source cannot be improved with the use of a lens, since this implies an increase of the intensity of the source. The actual computation of power, modes launched and launching efficiency into each mode, is straightforward, but can be lengthy. Using the same approach, the relaunching from one multimode fibre to another can be calculated.

4.8 Fibre/fibre longitudinal coupling

Light can be launched into another fibre or fibres through longitudinal coupling. When two fibre cores are side by side, the evanescent fields form a coupling mechanism between the fibres. The coupled-mode theory can be applied to show mode-to-mode power transfer between the fibres. This can also be dealt with as a composite 2-core guide with its own eigenmodes.

4.8.1 Biconical coupler

This special form of longitudinal coupling is not exactly longitudinal coupling. It is an important practical way of making a fibre/fibre coupler. Schematically, when two fibres from a pair of reverse tapers are placed alongside each other, the power from one flows to the other. The actual mechanism of power flow involves the radiation of the power from the core into the cladding, which now form the new waveguide. Owing to symmetry, the power in the cladding waveguide couples naturally back to the reverse tapered core, and thereby completes the coupling action.

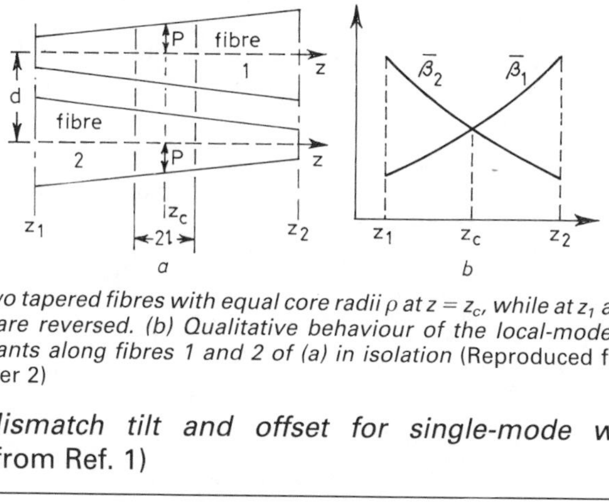

Fig. 4.31 *(a) Two tapered fibres with equal core radii ρ at $z = z_c$, while at z_1 and z_2 the core radii are reversed. (b) Qualitative behaviour of the local-mode propagation constants along fibres 1 and 2 of (a) in isolation* (Reproduced from Ref. 1 of Chapter 2)

Table 4.2: *Mismatch tilt and offset for single-mode waveguides* (Reproduced from Ref. 1)

	Gaussian beam
On axis	$\dfrac{P_0}{P_i} = 4\dfrac{r_0^2\rho_s^2}{(r_0^2+\rho_s^2)^2}$
Tilted at angle θ_i to axis	$\dfrac{P_0}{P_i} = 4\dfrac{r_0^2\rho_s^2}{(r_0^2+\rho_s^2)^2}\exp\left\{-\dfrac{(kn_i\theta_i r_0\rho_s)^2}{r_0^2+\rho_s^2}\right\}$
Offset r_d from axis	$\dfrac{P_0}{P_i} = 4\dfrac{r_0^2\rho_s^2}{(r_0^2+\rho_s^2)^2}\exp\left\{-\dfrac{r_d^2}{r_0^2+\rho_s^2}\right\}$

4.9 References

1 BARNOSKI, M. K., and PERSONICK, S. D.: 'Measurements in fiber optics', *Proc. IEEE*, 1978, **66,** pp. 429–441

2 KAO, K. C., and HOCKHAM, G. A.: 'Dielectric flow surface waveguides for optical frequencies', *ibid.*, 1966, **117,** 1151–1158

3 MILLER, S. E., MARCATILI, E. A. J., and TINGYE, LI: 'Research toward optical fibre transmission systems', *Proc. IEEE*, 1973, **61,** pp. 1703–1751

4 BLACK, P. W.: 'Fabrication of optical fibre waveguides', *Electrical Comm.*, 1976, **51,** pp. 4–11

5 TANAKA, S., INADA, K., AKIMOTO, T., and KOJIINA, M.: 'Silicone clad fused silica cover fibre, *Electron Lett.*, 1975, **11** (7)

6 OSANAI, H., SHIODA, T., MARIYAMA, T., ARAKI, S., HORIGUCHI, M., IZAWA, T., and TAKATA, H.: Effect of dopants on transmission loss of low-OH-content optical fibers', *Electron. Lett.*, 1976, **2,** pp. 549–550

7 MALITSON, I.H.: *J. Opt. Soc. Amer.*, 1965, **55,** pp. 1205–1209

8 FLEMING, J. W.: *Electron Lett.*, 1978, **14,** p. 326

9 OLSHANSKY, R., and KECK, D.: 'Pulse in broadening in graded index optical fibres', *Applied Opt.*, 1976, **15,** pp. 483–91

10 KEISER, G.: 'Optical fibre communications' (McGraw-Hill 1983)

11 MILLER, S. E., and CHYNOWETH, A. G. (Eds): 'Optical fibre telecommunications (Academic Press, 1979)

12 BASCH, E. E. (Ed.): Optical fibre transmission (Howard W. Sams, 1987)

Chapter 5

Material properties

The physical properties of the optical fibre are governed by the characteristics of its constituent materials and how the fibre is made and protected. A broad range of disciplines in physics, chemistry and material sciences are involved. In the course of the development of the fibre, in the past and now for the future, the depth of knowledge of these basic sciences was often found to be inadequate. The pursuit of the ultimate fibre has forced new advances to be made on all fronts. In this Chapter, discussion is aimed at demonstrating the scientific basis underlying each optical and mechanical property which the practical fibre has to meet. It will readily be seen that both theoretical and experimental methods and analytical instrumentation must be advanced step by step, often in sequence, in order to make significant progress.

Optical losses and dispersion, and their relation to material composition and structure, govern the optical properties. These have been discussed at some length in Chapter 4, and will be further analysed to illustrate particularly the purification problems. Mechanical strength, and its dependence on external physical and chemical environment, is closely related to the properties of brittle material and the fatigue-failure mechanism. The basic strength of the fibre and fibre durability will be treated at some length as an illustration of the interplay of mechanical, chemical and internal forces within the material system. It is an interesting demonstration of the interdisciplinary nature of science. The passivation of fibre surfaces opens a new field of surface chemistry. This is an extension of the study of the mechanical strength and durability of fibre.

The study of radiation-induced loss opens a difficult, and still far from completed, question of how to achieve better understanding of the effect of radiation on fibre. It reminds us that glass is a poorly understood but extremely important class of materials.

The nonlinear properties result from the interaction of the material with the incident electromagnetic wave. The electron and lattice interaction with photons is an exciting new area of research, made even richer recently owing to our ability to tailor the material structure and to generate highly coherent photons. In the fibre, owing to the confinement of optical energy within the small core region, the nonlinearity effects can build up over a long length so that nonlinearity can be induced even at a power as low as 1 mW. This has deleterious effects in limiting the maximum power which a fibre can carry, but it opens up possibilities for creating dispersion-compensated propagation and bistable components to serve as the basis for an optical switch.

5.1 Material losses

As discussed in the previous Chapter, the basic losses for material are due to the electronic absorption bands in the ultra-violet region and the molecular-vibrational absorptions in the infra-red region, including impurity absorptions particularly from OH^- ions, together with scattering due to small local inhomogeneities.

The choice of the right material system for optical-fibre waveguides was relatively arbitrary. The early intention was to choose a material with low loss in the optical spectral range, which is approximately 0·5—1 μm. This matched the wavelengths of the prospective light sources, as well as being in the spectral range where relatively transparent materials were known. The excursion into the ultra-violet and longer infra-red region was not attractive, since few transparent materials were known.

The rationales for the choice of high-temperature glass instead of crystalline materials and low-temperature polymers are interesting. Single crystals must be grown slowly. On a statistical basis, single crystals cannot be grown without some defects. Polycrystals contains many scattering centres. Hence, single-crystal fibre is a second best. Polymeric materials are built on C-H bonds and have relatively large molecules, which give rise to high scattering loss as well as broad absorption bands in the infra-red region. Only polymethyl methacrylate was of interest owing to its short polymeric chain length, and hence relatively low scattering, and its transparency in the optical-wavelength range. Indeed, high-purity fibre made with this polymer has been produced commercially with loss as low as 10 dB/km at 0·8 μm wavelength.

High-temperature glass has more likely prospects of reaching a better performance. Furthermore, the mechanical characteristics also appear to be superior. The history of development of fibre has supported this contention so far. In studying fibre loss, the characteristics of the glasses were determined in many ways. The basic ultra-violet and infra-red absorption were measured using standard optical spectroscopy. It was soon evident that the ultra-violet absorption edge was masked by impurity absorptions. Measurement on pure fused quartz later showed that spectral absorption can be identified from electronic levels and transitions observed for amorphous SiO_2. (Fig. 5.1)

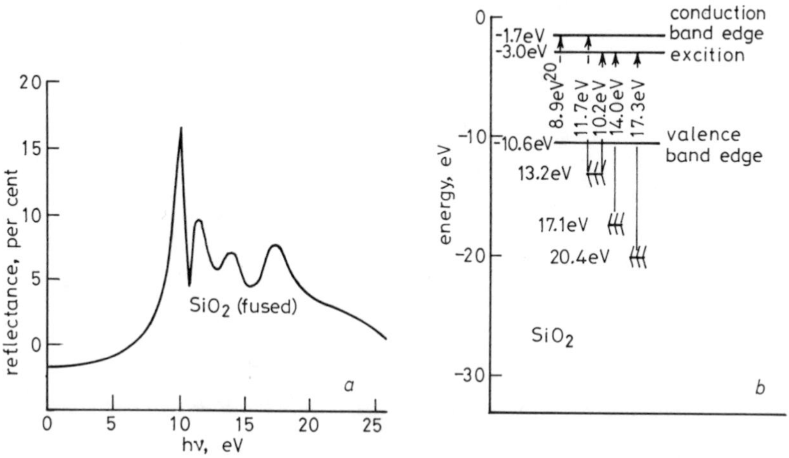

Fig. 5.1 *(a) Spectral dependence of the reflectance of fused quartz* (Reproduced from Ref. 1)
(b) Part of the experimentally observed electron energy levels and transitions for amorphous SiO_2 (Reproduced from Ref. 2)

The infra-red vibrational absorption has also been identified as due to Si–O–Si stretch vibrations when the O atoms move out of phase with their Si neighbours, and also as due to bending motion caused by O atoms moving away from the Si–Si line. Rocking motion has also been identified. These vibrations interact to provide several absorption peaks, and result in the infra-red transmission edge. (Fig. 5.2 and Fig. 5.3)

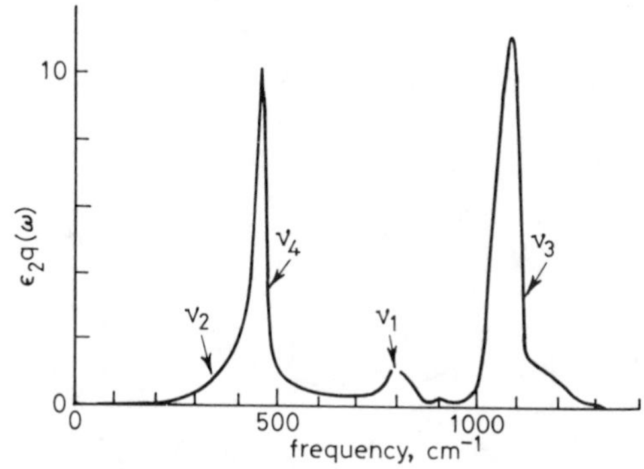

Fig. 5.2 *Experimental values for the imaginary part of the dielectric constant for vitreous silica* (Reproduced from Ref. 3)

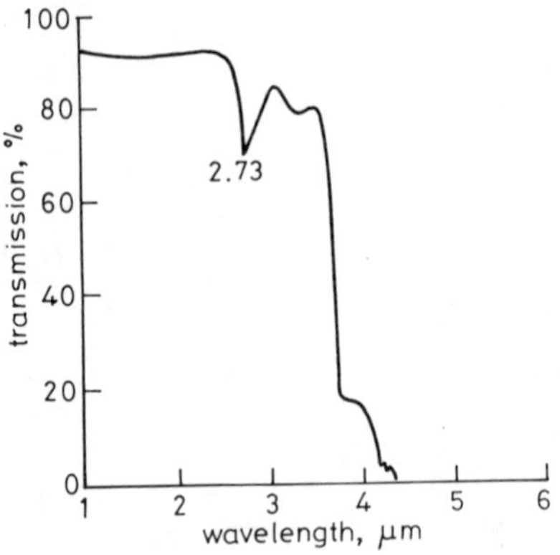

Fig. 5.3 *Transmission curve of commercial fused silica* (Reproduced from Ref. 4)

Spectroscopic studies also can be effectively used to study impurity absorption bands (Fig. 5.4). It was found that impurity absorption bands result from electronic transitions within the *d*-shell, known as ligand field transitions. The actual absorption coefficient of transition metal ions depends on how the given ion interacts with the ligand field of the host material and the associated charge-transfer processes. Even though this work became less important, owing to the success in impurity-ion removal through purification, the knowledge gained is new to material science and important in understanding how internal fields interact with isolated impurities.

Water absorption has been carefully studied. Two basic vibrational modes have been identified at 2·73 μm (Fig. 5.3) and 6·25 μm, corresponding to

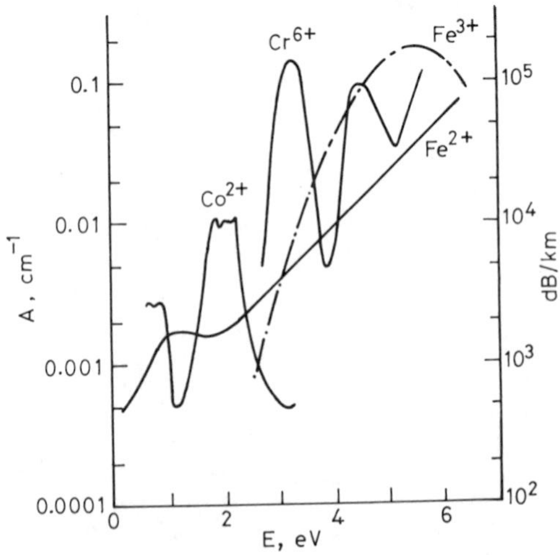

Fig. 5.4 *Absorption of some ligand field and charge-transfer transitions in glass* (Reproduced from Ref. 5)

stretching and bending. Coupling between them is due to the ease with which hydrogen attaches to the chain end of the Si–O bond. The coupling produced a large number of overtones and harmonics, (Fig. 5.5).

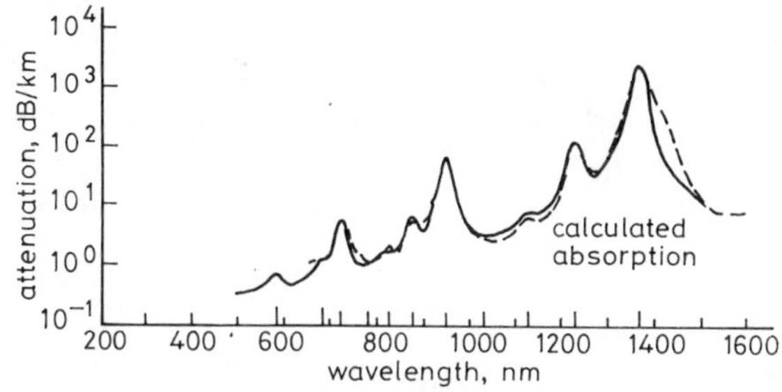

Fig. 5.5 *Calculated absorptive attenuations as a function of wavelength* (Reproduced from Ref. 6)

The task of eliminating impurities in glass systems, on one hand, gives rise to the need for impurity analysis both for the raw constituent materials and for the resulting glass, and, on the other hand, gives rise to the need to devise purification techniques and material-handling techniques that introduce as few additional impurities as possible. A list of analytical techniques suitable for detecting trace impurities down to a few parts per billion is shown in Table 5.1.

Table 5.1: *Techniques for trace and ultra-trace analysis* (Reproduced from Ref. 1)

Atomic-absorption spectroscopy
Atomic-fluorescence spectroscopy
Coulometric titration
Electron-capture gas–liquid chromatography
Electron-probe microanalysis
Emission spectroscopy
Fluorimetry
Ion-specific electrodes
Kinetic measurements
Mass spectroscopy
Neutron activation
Nuclear-track counting
Polarography
Radioisotope dilution
RF induction-coupled atomic emission
Stable isotope dilution
Spectrophotometry
Titrimetric methods
X-ray fluorescence

These techniques have their individual limitations. For example, depending on the host material, some impurities can be masked, or samples must be in the form of a solution, or the speed and cost of measurement may be too high for routine use. Nevertheless, a combination of several of these analytical techniques can provide an accurate determination of trace elements. Neutron-activation analysis (NAA) was considered the best technique for quantitative analysis, particularly for glasses. X-ray fluorescence, atomic-absorption and radioisotope techniques and infra-red spectroscopy were extensively used. NAA results on silica analysis are shown in Fig. 5.6 and Table 5.2.

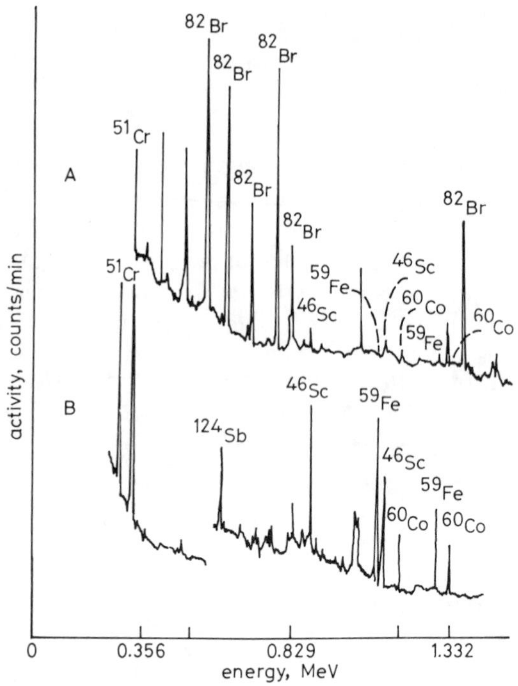

Fig. 5.6 *γ-ray spectra of irradiated KMC powdered silica* (Reproduced from Ref. 11 of Chapter 4)

Purification and handling must be tailored for individual cases. For compound glass making, the glass constituent material may be in oxide or carbonate forms. Carefully designed chemical purification processes have been evolved. In principle, pure reagents must be used and the process carried out in a non-contaminating environment, namely, in high-purity quartz containers and in a clean room or hood. The method of purification can be based on an ion-exchange technique in which a specific impurity ion is extracted by forming precipitates or by dissolving it in a specific solvent.

The chemical-vapour-deposition process was found to be particularly

Table 5.2: *Detection of trace elements in silica*[a] (Reproduced from Ref. 1)

Supplier	Concentration (μg/g)		
	Fe	Co	Cr
(1)	1·24	0·030	1·90
(2)	0·82	0·005	2·16
(3)	1·60	0·041	0·17
(4)	1·00	0·008	0·19
(5)	0·68	0·009	0·009
Suprasil	ND	ND	ND
	<0·1	<0·003	<0·005

[a] Semiquantitative γ-ray spectrometric surveys using synthetic comparison standards

effective as a method for producing glass-constituent materials in oxide form from volatile liquids containing the required elements. The purification is achieved through differential vapour pressure of the impurities. This will be discussed in detail in Chapter 6. It suffices to note that an important reagent for the CVD process is silicon tetrachloride. Infra-red spectroscopy of the commercial grade is shown in Fig. 5.7.

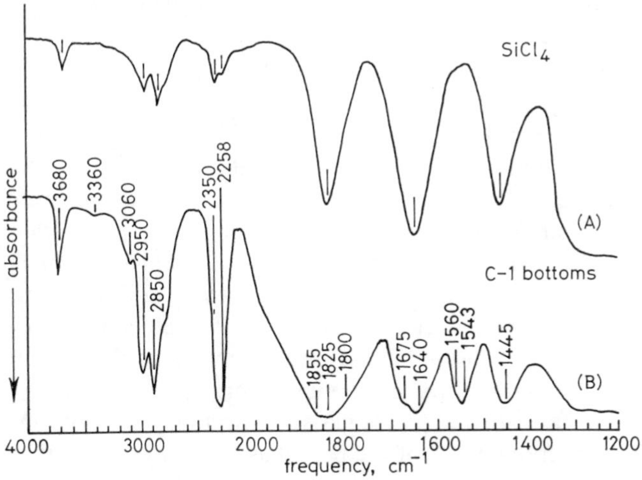

Fig. 5.7 *Infrared spectra of commercially available $SiCl_4$.* (Reproduced from Ref. 11 of Chapter 4)

Various absorption bands have been identified as due to OH groups, CH–Cl, H–Cl and Si–H bands, as well as $COCl_2$ absorption. Contamination of

CCl_4 and Si_2OCl_6 can also be seen. The contaminations are introduced because $SiCl_4$ is a good solvent for many organic materials, and is readily decomposed to form hydrolysis and decomposition products when exposed to air. Purification, however, poses no special problems.

Handling of pure materials is not difficult provided suitable precautions are taken. Exposure to the ambient atmosphere must be avoided. The containers should be made from high-purity material which sheds no particles and is not dissolved by the material stored in them. It is also important to understand temperature effects, electrolytic effects and hydrolysis effects. The experience of making high-purity materials has greatly increased our chemical-analytical skill, as well as spectroscopic-analysis methods. The benefit is felt not only in enabling low-loss fibres to be made, but also in a broad range of other chemical and material-science related areas.

5.2 Scattering loss

Scattering-loss mechanisms include particle inclusions, phase separation, devitrification, random fluctuations in glass, dislocation and inclusions in crystals. Single crystals have a regular structure. If perfect, scattering loss is absent. However, all crystals have imperfections such as crystal structural dislocations and polycrystalline tendencies. Scattering will occur when light waves encounter such discontinuities, whose dimensions often are large compared with light wavelengths. Glass-scattering loss, which has been discussed earlier, is due to the small random fluctuations corresponding to Brownian motion within a liquid (see Chapter 4). Here we will extend the discussion to include the material composition and structure-related effects.

Glass formation proceeds thermodynamically. The possibility of conditions favouring crystallisation is always present during some phase of the thermal cycle. Crystallisation is the process of formation of microcrystalline regions. This can occur most readily if glass is subjected to high temperature over a long period near, but below T_g, the fictive temperature, or glass-transition temperature. Control of crystal growth in fibre is enabled through tailoring the the thermal cycle so that the glass is cooled slowly to T_g and then rapidly after T_g. The slow cooling is required in order to achieve good homogeneity.

Phase separation is another cause of inhomogeneity. In a multicomponent glass system the separation of the glass-forming component in different proportions to form two or more glass (amorphous) phases can occur. This occurs when the free energy of the polyphase system is lower than that of the single-phase system. In a liquid/liquid immiscibility situation, the kinetic barrier is small and the separation is limited only by diffusion. Phase separation is likely to take place to some degree in a multicomponent system. Rapid quenching will not prevent phase separation, but will reduce the spread of the phase-separated regions.

These further considerations are brought out to illustrate that tailoring of glass for optimum performance is a lengthy and complex problem. The scattering loss is expected to be low when using a relatively simple glass system. However, for more complex glasses the phase-separation and crystallisation effects become even more significant. Owing to the nature of the constituent elements the composition of glasses for minimum scattering loss is likely to be bounded in a narrow compositional range. Actually this situation demonstrates our lack of understanding of how to control material growth in glass form. Currently we are happy to leave it to nature to determine the resulting product, particularly for bulk-glass formation. However, as will be seen in Chapter 6, the chemical-vapour-deposition process opens up the possibility to force glass formation. This field, however, remains open for more in-depth research.

5.3 Other material-selection criteria

The selection of material is constrained by loss requirements. Other constraints are that the material must be coated by a second material of correct refractive index and compatible physical properties, in addition to being of low loss. Suitable material-dispersion characteristics for reducing waveguide dispersion would be an added advantage. Unless so compelled for special reasons, single-crystal materials are to be avoided. They are slow to make in long length and pose significant problems in finding a suitable second material for cladding. Besides, the scattering loss due to inclusions and dislocations is usually significant. For glass, the choices are relatively wide even under all the desired properties and constraints. It will be seen that a silica-based fibre has almost ideal characteristics. Hence the discussion will be concentrated on such glass systems.

5.4 Refractive-index issue for high SiO_2 glasses

Silica has a refractive index n_D of ~1·46 at sodium D-line wavelength. It is wavelength dependent (see Fig. 4.8). It can be altered by the addition of oxides of dopant materials such as germanium. The n_D for various dopant silica with oxide addition in mole % is shown in Fig. 5.8.

It is seen that most dopants increase the index, while boron and fluorine can lower the index. The refractive index is a direct manifestation of the interaction of the material system with photons. The mechanisms giving rise to the interaction and frequency dependence have been phenomenologically studied in solid-state physics by designating a polarisability parameter P associated with the material. It is intimately linked with the vibrational modes and electronic transitions of the material system. A few mole % of oxide

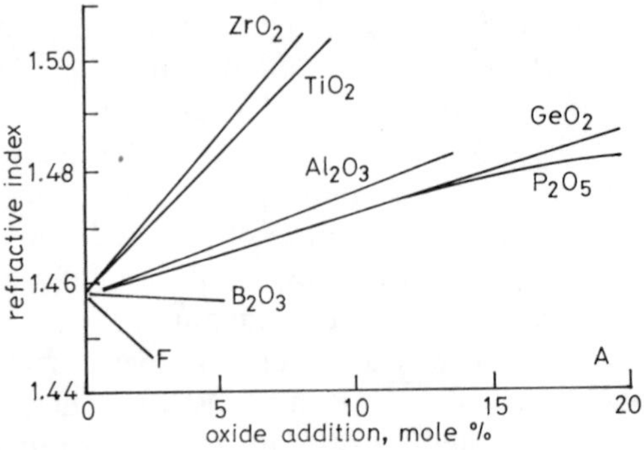

Fig. 5.8 *Refractive index n_D of doped silica glasses, ZrO_2, TiO_2, Al_2O_3* (Reproduced from Refs. 7–11)

addition is often sufficient to produce the desired index change for fibre-waveguide applications.

If SiO_2, owing to its extreme purity, is to be used as the fibre core, then fluorine-doped SiO_2 cladding must be used. If a little addition of dopant such as GeO_2 is used with SiO_2 to form the core material, then SiO_2 itself can be the cladding material. The choice is, in fact, dependent on other considerations such as cost, ease of fabrication and mechanical properties.

5.5 Expansion coefficients

The physical integrity of a fibre is somewhat dependent on the expansion coefficients of the core and cladding material. For a single-material homogeneous fibre, the fibre is stress free. For a fibre with core and cladding the residual stress is the result of a physical-property mismatch between the two material systems. If the core has a higher coefficient of expansion and solidifies later than the cladding, the resulting fibre has a compressed cladding; this is a strong and stable structure. If the core material has a lower expansion coefficient or higher temperature of solidification than the cladding, the resulting structure will have the outer surface in tension. This will not be as strong a structure as the previous situation. In practical silica and doped-silica fibres, the cladding is much thicker than the core. Hence, the requirement on matching or choosing the best expansion-coefficient mismatch is not a serious issue. The reason lies in the mechanical strength of glass, which is very strong in compression but weaker in tension. This will be discussed in Section 5.7. Some expansion coefficients of doped SiO_2/glass are shown in Fig. 5.9.

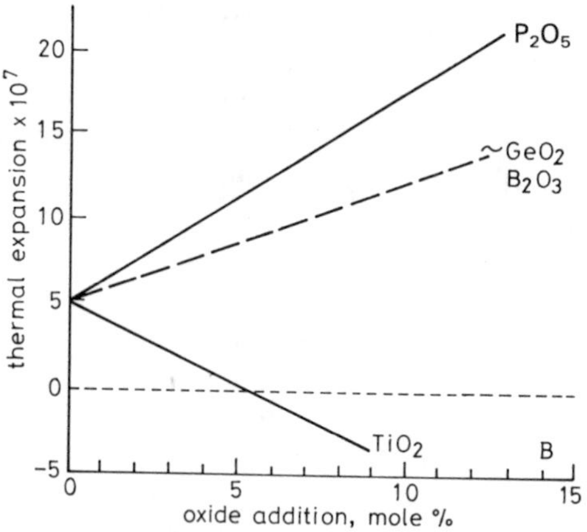

Fig. 5.9 *Thermal expansion (α) of doped silica glasses, P_2O_5* (Reproduced from Refs 9, 12, 13, 14)

5.6 Viscosity

The change of viscosity of glass with temperature affects the fibre-making process, but is seldom severe enough to prevent fibre formation. A viscosity/temperature relationship is shown in Fig. 5.10 which indicates that boron and phosphorus lower viscosity very significantly.

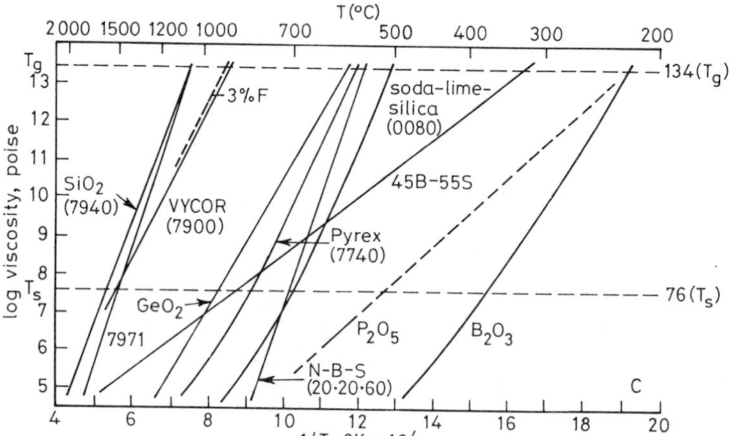

Fig. 5.10 *Viscosity of some waveguide glasses and dopants* (Reproduced from Refs 15, 16, 11, 17, 13, 18, 19, 20)

The doped SiO_2 glass and pure fused SiO_2 glass can be seen to be good for low-loss fibres, and the mechanical integrity is excellent. The strength and durability will be shown to be exceptionally good, except at very high temperature and in corrosive environments. It will also be clear that glass science is fundamentally an uncharted area closely linked with the basic dynamics of glass formation. The lack of understanding does not prevent glass being used extensively to meet many application requirements. But, as will be seen from fibre studies and other solid-state-physics developments, the opportunity to make glass do more in the service of man is very promising.

5.7 Fibre mechanical strength and durability

Glass is a brittle solid. Its properties in bulk or fibre forms are basically similar, although the residue stress distribution could make its mechanical properties apparently different.

The strength of fused silica is based on the Si–O bond strength. A classical model postulates that bonds between atoms can be described by a repellent force and an attractive force contributed by the basic nature of atomic bonds. (Fig. 5.11) Both are r^{-m} dependent with $m = m_1$ for attraction and $m = m_2$ for repulsion. The resulting force holds the bond together. When external force is applied, compressive forces are less likely to modify the force balance significantly unless shear is present, while tensile forces could break the bond if it exceeds the maximum bond strength. Over a certain force range, the bond will stretch elastically to compensate for the external force. This is the elastic region. For pristine glass this range can be as high as 20%.

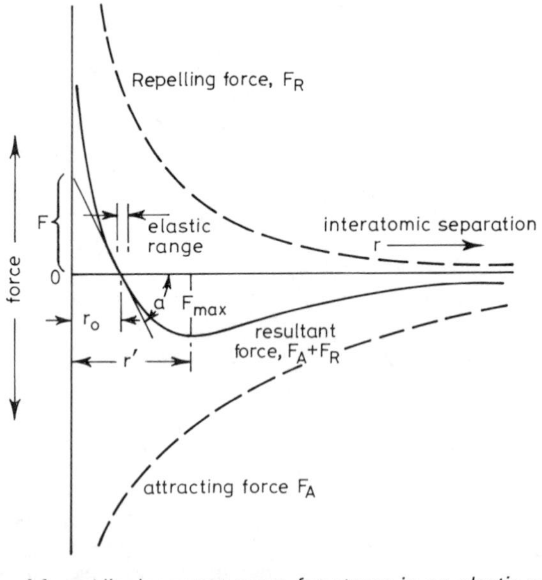

Fig. 5.11 *Schematic of force/displacement curve for atoms in an elastic solid* (Reproduced from Ref. 11 of Chapter 4)

The stress/strain relationship over a linear range is known as E, Young's modulus:

$$E = \frac{\text{stress}}{\text{strain}} = \frac{d\sigma}{d\varepsilon} \quad (5.1)$$

The corresponding shear modulus G and bulk modulus K are given by

$$G = \frac{E}{2(1+v)} \text{ and } K = \frac{E}{3(1-2v)} \quad (5.2)$$

where v is the Poisson ratio, which expresses the ratio of transverse shrinkage and longitudinal elongation. Some typical elastic constants are shown in Table 5.3. For fused silica the Young's modulus remain linear over 10% extension.

Table 5.3: *Typical values of elastic constants for glasses*

	E (GPa)	G (GPa)	v	K (GPa)
Fused silica fibre*	71·9	31·5	0·14	33·4
Aluminosilicate†	91	36·1	0·26	63·2
Borosilicate (Pyrex)†	61	25·0	0·22	36·3
Soda–lime–silica†	74	30·6	0·21	42·5
Lead–silicate†	61	25·2	0·21	35·1

* From Ref. 21
† From Ref. 22

The ultimate bond strength for SiO_2 is estimated to be 20 GPa stress or $\sim 5 \times 10^6$ lbf/in^2 (10^6 lbf/in^2 = 7 GN/m^2 = 7 GPa = 700 kg/mm^2), which is the work done per unit cross-sectional area of the solid as it is being pulled apart by the tensile force. It can also be interpreted as the energy needed to form the new surfaces. The theoretical cohesive strength σ_t can be expressed as

$$\sigma_t^2 = \frac{2\gamma E}{\pi a} \quad (5.3)$$

where γ is the surface energy of the material. This bond strength is for the Si–O bond with a bond distance a = 1·6 Å.

The strength of practical glass in pristine state, and not being influenced by reagents in its environment, should attain the theoretical cohesive strength. Most glasses, however, always operate in an unprotected environment, and therefore become less strong as a result of the glass network being replaced by weaker bonds formed with external reagents. The commonest is the formation of the silanol group SiOH. The bonds involved in silanol-group compounds have cohesive strengths much lower than the Si–O bond, and are formed when the Si–O network interacts with OH^- ions.

Another mode of failure is due to the local stress concentration exceeding the cohesive strength of the material. The local stress is magnified along flaws or cracks, particularly those transverse to the applied stress. The stress at the crack tip can reach failure stress before the average stress reaches the same levels. In addition, when the glass network is under stress, the stress provides enough energy to form new bonds with reagents within the vicinity. This will result in a weaker bond with a cohesive strength lower than the stress concentration. Hence the bond will break and the crack deepens. The stress concentration at the tip increases as a result, and the chance of further new weak bond formations also increases. Progressive crack propagation continues until the crack-tip stress exceeds the cohesive stress of the glass network and total fracture occurs. This progressive failure mode is called fatigue. It is a stress-induced mode of failure when corrosive reactants are allowed to reach the crack tips.

Initially all these failure mechanisms were not well understood. The low glass strength was regarded as due to the existence of flaws which caused local concentration of stress. Several theories were proposed. One controversal theory due to Cox assumes that a certain fraction of the Si–O bonds in a glass are in a broken state at any given moment as a result of the statistical spread of vibrational energy. Since a broken bond is incapable of supporting a stress, an extra load will momentarily be placed on the bonds that are in the vicinity of the broken bond when a stress is applied. If a critical number of neighbouring bonds are simultaneously broken, a fracture will propagate. Time-dependent, temperature-dependent, and moisture-dependent effects can all be predicted.

Another well accepted theory is due to Griffith. This theory starts by assuming that glass has flaws, and concerns itself with the condition under which the flaw will propagate. It is assumed that the flaws have the geometrical shape of narrow cracks, with small radii of the curvature at their tips where applied stresses are concentrated. The stress at the tips can be calculated from elasticity theory. The calculation has been carried out for a straight crack of elliptical cross-section with the applied stress perpendicular to the crack, as shown (Fig. 5.12).

The stress σ at the crack tip in the direction of applied stress S is

$$\sigma = S\left(1 + \frac{2L}{a}\right) \tag{5.4}$$

where L is the crack length (semi-major axis of the ellipse) and a is half the crack width (semi-minor axis of the ellipse). With the radius of curvature $\rho = a^2/L$ at the crack tip and assuming $L \gg a$,

$$\sigma = 2S\left(\frac{L}{\rho}\right)^{\frac{1}{2}} \tag{5.5}$$

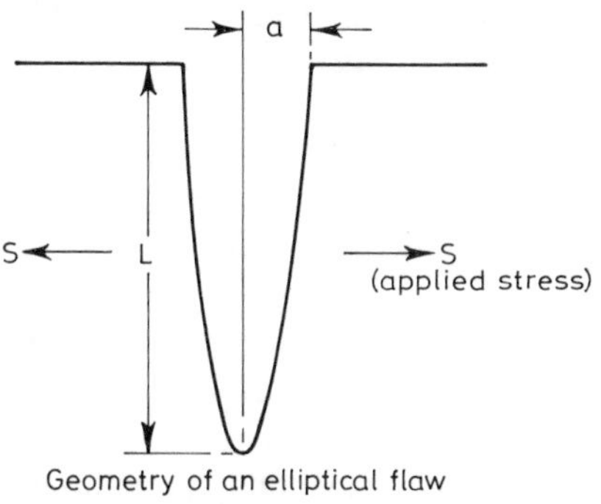

Fig. 5.12 *Calculation of stress at tips for straight crack of elliptical cross-section* (Reproduced from Ref. 23)

The fracture stress σ is inversely proportional to the square root of the crack length. This dependence was found by Griffith in his well-known equation

$$\sigma = \left(\frac{2E\gamma}{\pi L}\right)^{\frac{1}{2}} \tag{5.6}$$

which gives the fracture stress for a crack of elliptical shape with semi-major-axis length L, and where E is Young's modulus and γ is the surface energy. The stress at the crack tip with load exerted along the minor axis of the ellipse is equal to the cohesive or bond strength. For fibres with σ in the region of GN/m^2, the crack size responsible is of the order of 100 Å. This theory is important since it describes the stress at the crack tip if the surface contains flaws.

In a practical fibre the existence of surface flaws is a statistically rare event, but is statistically always present with certain probability over long fibre lengths. The cause of the flaw could be particle inclusion during any part of the fibre-making process, or mechanical damage subsequent to fibre formation. Furthermore, under high stress, stress-induced flaw formation could occur anywhere where local stress is marginally higher than in the neighbouring region. A statistical strength variation occurs along the fibre length. The strength of the entire fibre length must be expressed as the stress of the weakest point even if fatigue is not present. Hence, fibre strength must be expressed by the use of such a statistical distribution, which describes the distribution of rare events. One popular distribution is the Weibull distribution. This type of study is known as fracture mechanics—a field popularised after the discovery of time-dependent fatigue mechanisms leading to failure.

Fibre strength as a function of flaw distribution can be measured by testing long lengths of fibre. It is dependent on fabrication conditions and test conditions. In a controlled manufacturing environment where the fibre-making process is designed to avoid particle inclusion and accidental mechanical damage, and where the fibre is immediately coated with a protective non-abrasive coating, the statistical-strength data can help towards establishing fibre-service durability.

5.8 Time-dependent failures

The major reagent which readily attacks the Si–O network is OH^- derived from water or any chemical with OH^- ions. Molecular hydrogen has not been found to substantially influence fibre strength even though diffusion of hydrogen in the fibre does take place. OH^- attaches to glass surfaces in at least three general ways, which have been identified spectroscopically.

The surface structure of oxide glasses depends mainly on the reaction of nonbridging oxygen at the surface during surface formation. These bonds may react rapidly with atmospheric water to form silanol (SiOH) groups. The surface of a glass is therefore normally not free from metal-hydroxyl groups. The structural arrangements depend on glass composition, thermal history, humidity and surface treatment after melting and cooling.

The presence of physically absorbed molecular water is indicated by the presence of broad absorption bands at about $3450\,cm^{-1}$ and another at about $1250\,cm^{-1}$. Other bands indicate that some hydroxyl groups are sufficiently close together to be hydrogen-bonded.

Typical test data are shown in Figs. 5.13 and 5.14.

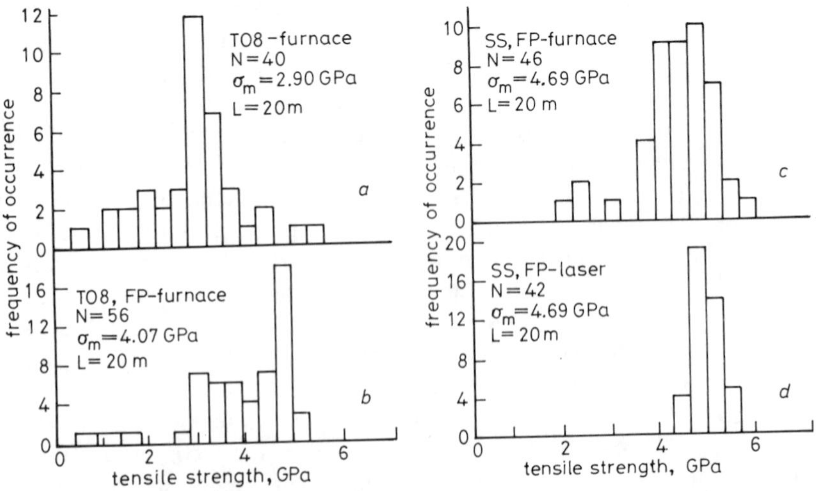

Fig. 5.13 *Fibre strength distribution as a function of drawing and preform condition* (Reproduced from 24)

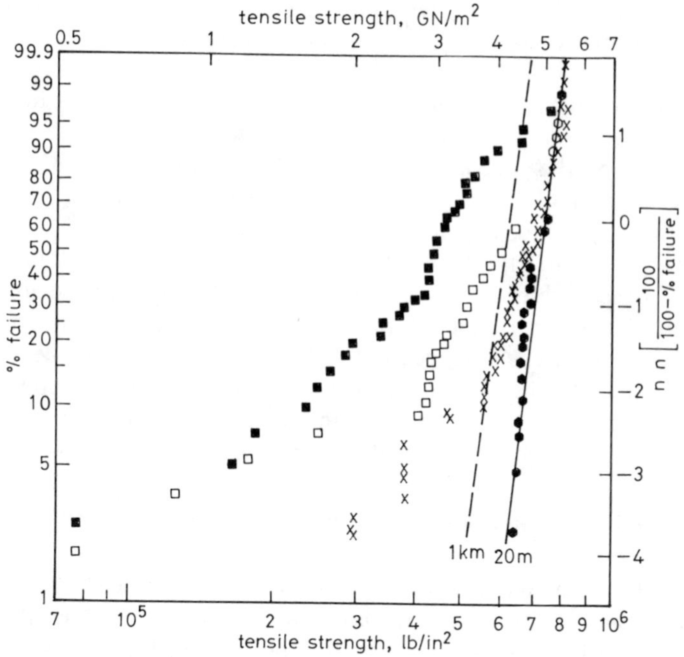

Fig. 5.14 *Cumulative failure probability plot of data* (Reproduced from Ref. 24)

The density of isolated SiOH groups on a silica surface has been calculated as 1·4 groups per 100 Å and that of hydrogen-bonded groups as 3.2 groups per 100 Å. Fig. 5.15 indicates the types of hydroxyl groups existing on silica surfaces. These are isolated silanol groups (Fig. 5.15*a*), two OH groups on one silica atom (*b*), and adjacent OH groups bonded with hydrogen bond and molecularly absorbed water (*c*). The relative concentration of each group on the silica surfaces depends on the melt atmosphere and thermal history of the glass, and the temperature and humidity during the spectroscopic measurement.

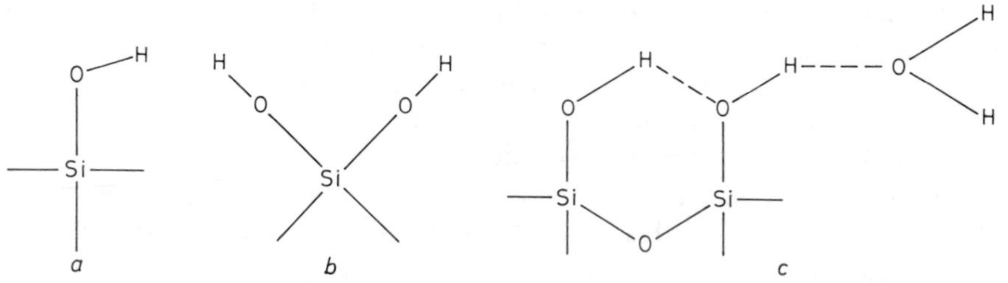

Fig. 5.15 *Hydroxl groups found on silica surfaces (a) isolated OH groups; (b) two OH groups on one silica atom; (c) hydrogen-bonded groups with an adsorbed water molecule* (Reproduced from Ref. 23)

Internal silanol groups near the surface have also been observed. These internal groups do not form at room temperature because of the low diffusion coefficient of water in bulk silica, but can take place at elevated temperatures. The silanol bonds are responsible for most of the lowering of fibre strength. This discussion is related to a silica surface; but it is applicable to other glass surfaces, since the mechanisms responsible for weakening the other glass surfaces work in a similar fashion. However, the silica surface is the strongest and most resistant to reagents.

A theory due to Hilling and Charles explains delayed fracture in glass in terms of stress-induced corrosion by water at flaw tips. Crack propagation is assumed to be an activated process in which the activation energy is stress-dependent. The basic equation for crack velocity can be developed from the absolute rate theory of chemical reactions. If the rate-limiting step for crack propagation is assumed to be an attack of the silicon–oxygen bonds by hydroxyl ions, the crack-propagation equation is

$$V = V_0[\mathrm{OH}^-][A_g] \exp \frac{-\Delta E_{\ddagger} + \sigma \Delta V_{\ddagger}/3 - V_m \gamma/\rho}{RT} \tag{5.7}$$

where OH^- is the hydroxyl-ion activity at the crack surface and $|A_g|$ represents the chemical activity of a flat glass surface in contact with the corrosive environment. The first term in the exponential $\Delta E_{\ddagger}$ represents the activation energy of the chemical reaction. The stress at the glass–liquid interface is given by σ, and the activation volume for the chemical reaction is represented by $\Delta V_{\ddagger}$. The final term in the exponential accounts for the changing chemical activity of the glass surface with surface curvature. The molar volume of the glass is given by V_m, γ is the interfacial surface tension at the glass–medium interface, and ρ is the radius of curvature of the glass surface. This is sometimes written in an alternative form:

$$V = V_0 \exp(\beta\sigma) \tag{5.8}$$

Where V is the velocity at which the glass surface corrodes under a tensile stress, V_0 is the rate of corrosion with no stress, and β is a constant parameter.

Experiments on the velocity of crack propagation at relatively low stresses verify this exponential form. For a given system (environment, temperature and glass composition), there is a unique relation between the crack velocity V and the crack-tip stress-intensity factor K_I. However, at higher stresses the velocity of crack propagation assumes different forms. Typically a trimodal curve is obtained. Two such curves are shown in Figs. 5.16 and 5.17.

At lower stresses (region I in Fig. 5.16) the velocity is exponentially dependent on stress. In this region the crack velocity can be expressed as a power function of the stress-intensity factor:

$$V = AK_I^n \tag{5.8}$$

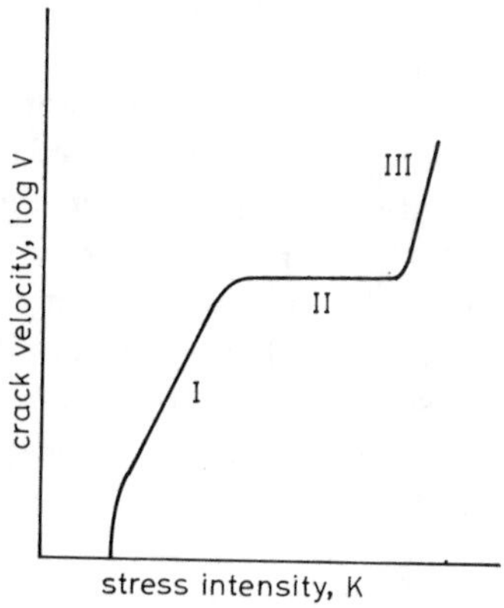

Fig. 5.16 *Trimodal characteristics of crack propagation* (Reproduced from Ref. 23)

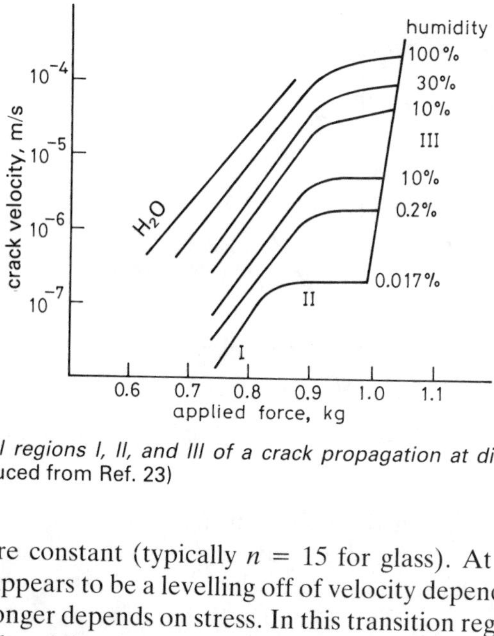

Fig. 5.17 *Trimodal regions I, II, and III of a crack propagation at different humidities* (Reproduced from Ref. 23)

where n and A are constant (typically $n = 15$ for glass). At higher stresses (region II) there appears to be a levelling off of velocity dependence on stress, until velocity no longer depends on stress. In this transition region the velocity still depends on humidity. It is suggested that, in this region, the crack-propagation velocity is limited by the rate of transport of water vapour to the crack tip. At still higher stresses (region III) the velocity again increases exponentially with stress, and is no longer dependent on the environment. For most glass systems, regions II and III occur at high velocities, and the crack-propagation time in region I governs the time delay to failure.

For low-stress failures, the fracture surface of glass has certain characteristics. The radial region closest to the initiating flaw is smooth and is known as the 'mirror region'. Beyond this there is a misty region where the surface roughens, and further away is a region of gross roughening known as 'hackle'. The small channels in the mist are the beginnings of branching of the crack, and the hackles are more extensive branchings of the crack.

By measuring the temperature dependence of crack propagation, there is confirmation that the activation energy of the chemical process involved is that of the formation of Si–O–H bond influenced by the energy supplied by the stress. A typical result is shown in Fig. 5.18. In order to predict the time to failure of a fibre, the stress/time relationship the fibre is expected to undergo must be known. For constant stress and a known maximum flaw size, the duration before failure can be predicted from experimental data for a fibre with pristine surface and is as shown in Fig. 5.19.

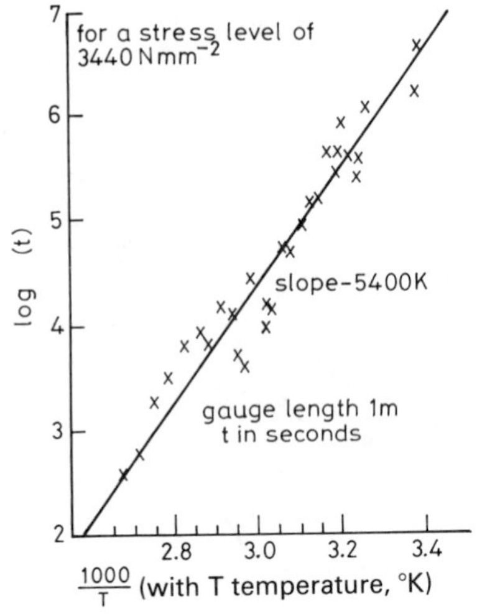

Fig. 5.18 *Effect of temperature on time to failure (Reproduced from Ref. 23)*

In order to establish the maximum flaw size along the entire fibre, it is possible to pre-stress the fibre at a stress generally called the proof stress, so that flaw sizes larger than those which can withstand the proof stress are eliminated. This conclusion is valid only if, during proof testing, minimal flaw propagation has taken place. This means that, in a proof test, the unloading of stress must be done over a very short time period and in a dry environment. A preferred time for proof testing is immediately after fabrication. Theoretically it is possible to correct for crack propagation by converting the proof stress to an after-proof test stress. Typical estimated values are shown in Fig. 5.20.

Proof testing can be done with constant stress or strain. After proof test the time to failure can be estimated for constant service stress. Such predictions are guaranteed only if the proof-test condition and all material constants are sufficiently and accurately known for a valid extrapolation over many orders of

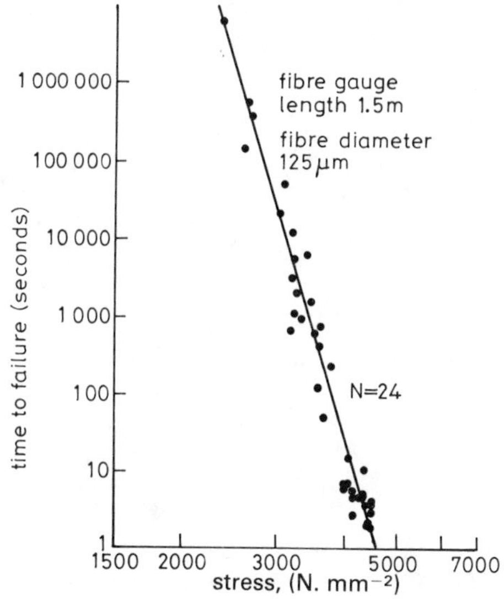

Fig. 5.19 *Stress-induced time to failure in air of a fibre* (Reproduced from Ref. 23)

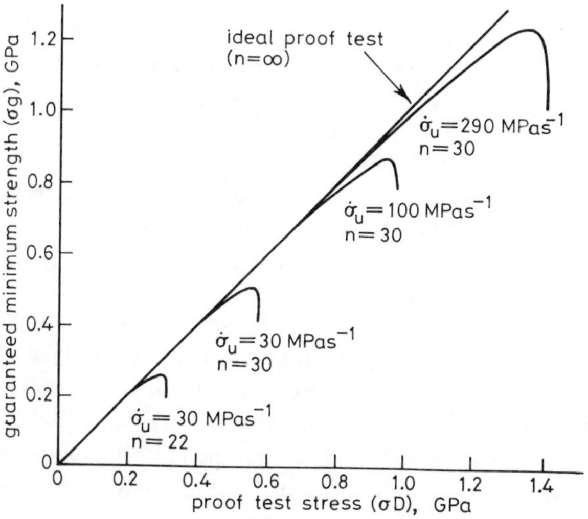

Fig. 5.20 *Effect of crack growth during proof testing on guaranteed minimum strength* (Reproduced from Ref. 25)

magnitude of time. Furthermore, the assumption of a power-law relationship for time-dependence failure on stress may not be sufficiently accurate to permit the extrapolation. Given those reservations, a life-prediction result is shown in Fig. 5.21.

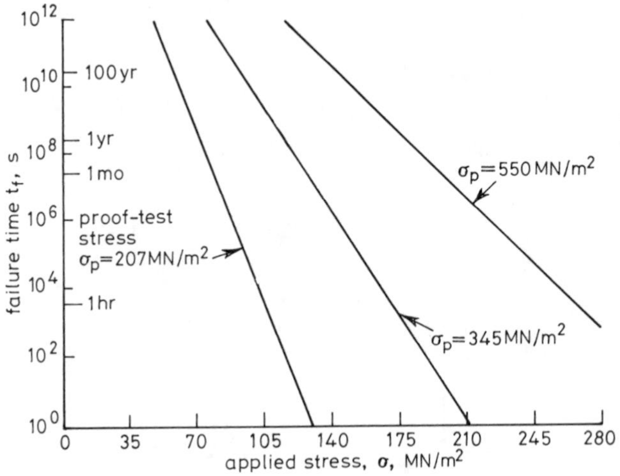

Fig. 5.21 *Life-prediction diagram for optical fibres based on proof-test stress* (Reproduced from Ref. 26)

The major issue facing failure prediction is the understanding of how OH^- ions or other reactive ions attacks the Si–O network, and the law governing the flaw growth rate. Furthermore, the service stress is not necessarily constant. Considerable uncertainty remains, so that the service stress should be kept sufficiently low to allow for a safety margin. The transport of corrosive agents, and the variation of the shape of the fracture tip with time and stress, are probably important parameters which influence durability.

Surface modification to prevent fatigue is approached in two ways. The first is to ensure that the fibre is made with as few surface flaws as possible, and is coated with a protective polymeric coating as soon as the fibre is made. Polymers do not stop OH^- ions from reaching the fibre surface, but they can form a bond with the Si–O surface. The second is to coat the fibre with a hermetic coating. The formation of a hermetic coating witout damaging the Si–O surface is not well understood. So far, the application of a hermetic coat of materials such as SiN or metal causes the fibre strength to be lowered. This could imply that the surface layer is affected up to several 100 Å. On the other hand, several 100 Å of hermetic material appear to be adequate for ensuring hermeticity. More work is undoubtedly needed for further clarification.

5.9 Optical nonlinearity in fibres

When the fibre operates with light at high power density, optical nonlinear effects become significant. Optical nonlinearity is the consequence of the interaction of incident photons with the material system at the sub-molecular and sub-atomic levels. Hence, deep understanding of the physics involved must be treated at a quantum-mechanical level. In the case of optical fibre the first step is to characterise the nonlinear properties without establishing the precise reason as to how the nonlinear interaction takes place at the sub-atomic level. The nonlinearity can loosely be grouped into at least five types. The first type is the interaction of incident photon with optical phonons (molecular vibrations), known as stimulated Raman scattering. The second type is when photons interact with acoustic phonons (molecular mechanical vibrations), known as the stimulated Brillouin scattering. Both these scattering mechanisms result in a shift in photon wavelength, and are referred to as inelastic scattering. The third to fifth types are caused by the intensity-dependent changes of refractive index when photons interact with lattice electrons instantaneously. The third type is when the electron interacts directly with the photon while the material stays passive; parametric processes can take place under this situation. The fourth type involves possible displacement of nuclei and/or anharmonic response of bond electrons as a result of the presence of the photon field; heating also contributes to this. The fifth type involves the trapping of the perturbed electron, so that the refractive-index change is long lasting; this effect is known as the photorefractive effect.

All these effects, of course, occur in bulk material as well as in fibre. Optical nonlinearity of fibre differs from bulk material owing to the fact that interaction distance can be maintained over a long length of a fibre at significant power levels.

5.10 Stimulated Raman scattering (SRS)

This is a fast process. The incoming photon is scattered inelastically by the optical phonons. The scattered radiation appears both above and below the incoming frequency. The upper and lower bands are referred to as anti-Stokes and Stokes band, respectively. For SiO_2, Stokes bands are shown in Fig. 5.22. The Boltzmann distribution of excited states makes upward frequency shift much less possible.

Stimulated Raman scattering takes place when the number of Stokes photons is sufficiently large. It gives rise to the generation of new frequencies, which can be used as new sources. If a second photon stream at a Stokes frequency is injected along with the pump, it is amplified. The Raman gain is dependent on the pump intensity and a gain coefficient. If the incoming photon stream is the information carrier, the information is transferred to the Stokes frequencies. This could become an interfering signal if the fibre is carrying other photon carriers. At the same time, the energy of the information carrier is dissipated to the Stokes bands. The energy can couple to the propagating Stokes frequency in both the forward and backward directions. Hence SRS can, on the one hand, limit the fibre signal power and

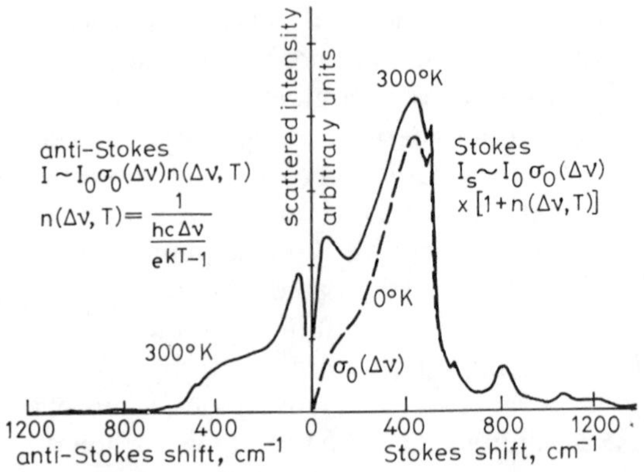

Fig. 5.22 *Stokes and anti-Stokes scattered intensities for fused silica at 0° and 300° K. n(Δν) is the phonon population factor*
h and k are the Planck and Boltzman constants, T is temperature in °K, c is the velocity of light and Δν is the frequency shift in cm^{-1} (Reproduced from Ref. 11 of Chapter 4)

set limits for multi-wavelength operation, but, on the other hand, it can serve as a means to obtain frequency-shifted tunable source and as an amplifier.

For a low-loss single-mode fibre with SiO_2 core doped with GeO_2, the signal power serves as the pump power and Stokes frequencies are generated and built up gradually along the length. When signal power starts at about 3 W it has been shown that, along a typical single-mode fibre, the Stokes power builds up to equal to the pump power and can occur over a distance of a few kilometres. The actual power and distance are dependent on the operating V value, which sets the maximum power density and wavelength of the signal as well as the waveguide loss. This sets a high operating power limit for fibre systems, and it is important to note that the nonlinear power-conversion efficiency is low until near critical power is reached.

If two or more carriers are present along a fibre the SRS can cause significant crosstalk, since the carriers are likely to be within the Stokes band. A worst-case analysis can be carried out for a 2-signal case, where one signal is regarded as the pump and the other as the power in the Stokes channel. If they start equal, the pump would suffer strong nonlinear attenuation, which, in turn, causes crosstalk in the Stokes channel. A typical power of 10 mW per channel could give rise to 20% crosstalk along a long fibre of about 100 km. Experimental data reported in actual laboratory systems gave −25 dB crosstalk for a 1 mW-power-per-channel system operating at 1·26 and 1·34 μm (Ref. 28), while the crosstalk between two channels in a 45 km system as a function of channel-number separation and power are as shown in Fig. 5.23.

For amplification and signal-generation applications, SRS promises to be useful. The Raman gain varies with the material. As the GeO_2 content is increased, Raman gain becomes larger. Pure GeO_2 fibre is an attractive

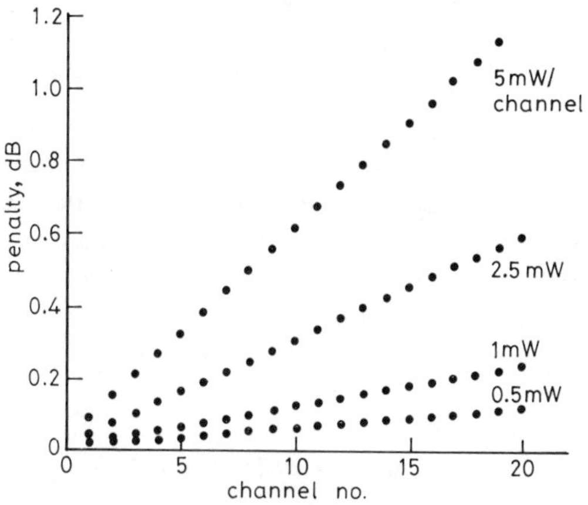

Fig. 5.23 *Power penalty on shortest-wavelength channel as a function of channel number for different but equal powers per channel* (Reproduced from Ref. 27)

Raman-amplifier medium. Doped SiO_2 with deuterium has also been found to increase the gain coefficient. A gain of 4000 for 1·56 μm wavelength has been demonstrated when the fibre was pumped at 1·06 μm along a 100 m D_2 doped fibre. (Ref. 29)

As a signal generator, the broad Stokes band can be used as a broad spectrum source for fibre characterisation covering the spectral region of 1—1·6 μm.

5.11 Stimulated Brillouin scattering (SBS)

This type of inelastic scattering is a maximum in the backward direction and zero in the forward direction owing to the quantum-mechanical selection rule. The phonon energies are small, resulting in small frequency shifts of up to 3×10^{10} Hz or 1 cm^{-1}. The gain $G = G_B(\Delta\phi_B/\Delta\phi_P)$, is much larger than SRS. It becomes significant if the pump line is sufficiently narrow. The Brillouin line width, is typically about 100 MHz. Thus Brillouin scattering can dominate when the signal source is near monochromatic, as used in coherent systems. A critical power for large conversion of pump power to Stokes frequency can be as low as 10 mW. This sets a significant limit to practical coherent systems, and it must be acknowledged and addressed. Methods to increase the critical power level at which SBS becomes significant, such as using a line-broadened source, are possible.

5.12 Refractive-index effects

When a single-mode fibre is pumped at ω_p, generating ω_s and ω_a as the two Stokes and anti-Stokes components, the conservation of energy demands

$$2\omega_p = \omega_s + \omega_a \text{ and } 2k_p = k_s + k_a \tag{5.9}$$

For

$$\delta k = k_s + k_a - 2k_p \neq 0 \tag{5.10}$$

This type of conversion will be efficient over a region when δk is small. This can be realised in several ways in a fibre. The dispersion between three modes in a multimode fibre can be chosen so that each mode carries one of the frequencies, and momentum conservation of phase match can be achieved. In the case of a birefringent fibre, phase match is again possible, while for a fibre with zero dispersion over a wide spectral region, phase match is always maintained. If ω_s and ω_p are supplied, the higher frequency ω_a can be generated.

5.12.1 Intensity-dependent refractive-index change

$$n = n_0 + n_1 I(\mathbf{r}, t) \tag{5.11}$$

where I is the time-average intensity, which is position dependent and time dependent. In a single-mode SiO_2 fibre, this effect is not observed. Self-focusing is normally induced if n_1 is positive.

5.12.2 Optical Kerr effect

This is an optically induced birefringence. A strong linearly polarised wave can induce birefringence in a fibre. Over a long length, the fibre offers phase difference between a second wave propagating along the slow and fast axis. This effect can be used to control the output polarisation of the second wave, and thereby acts as a polarisation modulator. Response time as short as 1 ps is possible.

5.12.3 Self phase modulation

A symmetrical-intensity pulse incident along a fibre whose refractive index is intensity dependent will cause the pulse centre to be retarded relative to the leading and trailing edges. In glasses the refractive index increases with intensity. Hence, the leading edge is down-shifted in frequency and the trailing edge is up-shifted in frequency. Defining the instantaneous frequency shift as

$$\delta\omega(t) = \frac{d\delta(\phi)}{dt} = -\frac{2\pi L}{\lambda}\frac{d\delta n}{dt}$$

a chirped pulse can be formed in a fibre by balancing the self-phase modulation and the linear dispersion. The leading edge of the input pulse is up-shifted in frequency while the trailing edge is down-shifted. The front edge of such a pulse, if allowed to pass through a delay line with negative group-velocity characteristics, will be slowed while the trailing edge will be accelerated, leading to a compressed pulse. A 3:1 compression from 90 fs to 30 fs has been achieved with this simple arrangement.

5.12.4 Soliton

If a fibre has negative group-velocity dispersion, a narrow pulse propagating through the fibre will maintain its shape since self-phase modulation and negative group delay compensate each other. A very narrow pulse then propagates with no change in shape over long distance. This condition is equivalent to satisfying the nonlinear Schrodinger equation, whose simplest non-trivial solution describes a soliton:

$$u(z,t) = \text{sech}(z)\exp(-it) \qquad (5.13)$$

5.12.5 Photorefractive effect

Photorefractive effects have not been observed in SiO_2 fibre, but they have been observed in a germania-doped SiO_2 fibre. Permanent refractive-index changes can be induced. This means that 4-way mixing can be established in a fibre to form self-conjugate reflectors.

5.13 References

1 PHILLIP, H. R.: 'Optical transitions in crystalline and fused quantity', *Solid State Commun.*,1966, **4,** p. 73

2 IBACH, H. and ROWE, J. E.: 'Electron orbital energies of oxygen adsorbed on silicon surfaces and of silicon dioxide', *Phys. Rev. B.*, 1974, **10,** p. 710

3 GASKELL, P. H. and JOHNSON, D. W.: 'The optical constants of quartz, vitreous, silica and neutron-irradiated vitreous silica', *J. Non. Crystal. Solids*, 1976, **20,** p. 171

4 ADAMS, R. V. and DOUGLAS, R. W.: 'Infrared studies in various samples of fused silica with special reference to the bands due to water', *J. Soc. Glas. Technol.*, 1959, **43,** p. 147T

5 KURKJIAN, C. R. and PETERSON, G. E.: 'Some material problems in the design of glass fibre optical waveguides', *Proc. Cairo 2nd Solid State Conf.*, 1973, **2,** p. 61

6 KECK, *et al.*: 'On the ultimate limit of low attenuate in glass optical waveguides', *Appl. Phys. Lett.*, 1973, **22,** p. 307

7 MAURER, R. D. and SCHULZ, P. C.: 'Optical waveguide'. Japanese Patent S46 6423, 1971

8 SCHULTZ, P. C.: 'Ultra violet absorption of titanium and germanium in fused silica' *Proc. 11th Int. Congr. Glass.*, 1977, **3,** p. 155

9 SCHULTZ, P. C.: 'Fused P_2O_5 type glasses'. US Patent 4 042 404: 1977

10 VAN UITERT, L. G.: *et al.*: 'Borosilicate glasses for fibre optical waveguides', *Mater. Res. Bull.*, 1973, **8,** p. 469

11 RAU, K., *et al.*: 'Progress in silica fibres with fluoride dopant', *2nd Top. Meet. Opt. Fib. Transm.*, 1977

12 RIEBLING, E. F.: 'Non ideal mixing in binary GEO_2–SiO_2 glasses', *J. Am. Ceram. Soc.*, 1968, **5,** p. 406

13 BRUCKNER, R. and NAVARRO, F.: 'Physikalisch-chemische Untersuchungen in System B_2O_3–SiO_2, *Glastech. Ber.*, 1966, **39,** p. 283

14 NORDBERG, M. E.: Glass having an expansion coefficient lower than that of silica'. US Patent 2 326 059: 1943

15 Corning Glassworks: 'Low-expansion materials', *Tech. Bull LEM*, 1969

16 Corning Glassworks: 'Properties of selected commercial glasses', *Bull.* 1949, **B-83**

17 ENGLISH, S. and TURNER, W. E. S.: 'The physical properties of boric oxide containing glasses and their bearing on the general problem of the constitution of glass', *J. Soc. Glass Technol.*, 1923, **7,** p. 155

18 CORMIA, R. L., *et al.*: 'Viscous flow and melt allotrophy of phosphorous', *J. Appl. Phys.*, 1963, **24,** p. 2245
19 NAPOLITANO, A., *et al.*: 'Viscosity determination of boron trioxide', *J. Am. Ceram. Soc.*, 1965, **48,** p. 613
20 KURKJIAN, C. R. and DOUGLAS, R. W.: 'The viscosity of glasses in the system Na_2O-GeO_2', *Phys. Chem. Glasses*, 1960, **1,** p. 19
21 MALLINDER, F. P. and PROCTOR, B. A.: 'Elastic constants of fused silica as a function of large tensile strain', *Phys. Chem. Glasses,* 1964, **1,** p. 91
22 HOLLOWAY, D. G.: 'Physical properties of glass' (Springer-Verlag, N.Y. 1973)
23 KAO, C. K.: 'Optical fibre systems' (McGraw Hill, 1982)
24 SCHONHORN *et al.*: 'Epoxy acrylate coated fused silica fibres with tensile strength >500 KSi (3·5 GN/m^2) in 1 km gauge lengths', *Appl. Phys. Lett.*, 1976, **29,** p. 712
25 TARIYAL, B. K. and KALISH, D.: 'Mechanical behaviour of optical fibres', *in* BRADT, R. C. *et al.* (Eds), 'Fracture mechanics of ceramics' (Plenum, 1978), Vol. 3, p. 161
26 TARIYAL, B. K. *et al.*: 'Proof testing of long length optical fibres for a communications cable', *Bull. Am. Ceramics Soc.*, 1977, **56,** p. 204
27 HEGARTY, J., *et al.*: 'Measurements of the Raman crosstalk at 1·5 μm in a wavelength-division-multiplexed transmission system, *Electron Lett.*, 1985, **24,** p. 395
28 TOMITA, A.: Crosstalk caused by stimulated Raman scattering in single-mode wavelength division multiplex systems, *Opt. & Quantum Electron,* 1984, **16,** p. 409
29 CHRAPLYVY, A. R., *et al.:* Optical gain exceeding 35αB at 1·56 μm due to stimulated Raman scattering by molecular D_2 in a solid silica optical fiber, *Opt. Lett.,* 1983, **8,** p. 415

Chapter 6

Fibre fabrication

Low-loss, high-bandwidth optical-fibre waveguides must be made with suitable materials to a physically easy-to-handle form; with near perfect geometry; with high strength and good flexibility; at low cost; and in long continuous lengths of many kilometres.

Historically, it was found that fibre can be drawn from molten glass with near-perfect cylindrical geometry. The surface-tension forces help it to attain a circular cross-sectional geometry. It was also noted that a fibre can be made by stretching a rod of larger cross-sectional dimension after the rod is softened by heat. Depending on the degree of softening, in other words depending on the viscosity, the geometry of the fibre can be a replica of the rod. When viscosity is high, the original cross-sectional shape of the rod is retained; while, when viscosity is low, the resulting fibre tends to have a circular cross-section. The uniformity of the fibre is governed by the constancy of the stretching rate, usually referred to as drawing rate. Furthermore, if the rod has a geometrical core/cladding arrangement, the fibre follows the geometry consistently, retaining, to all intents and purposes, the original ratios. Fibre can therefore be made from molten glass in crucibles, or from preforms, as the large-diameter rods consisting of core and cladding are called. Extensive work on fibre fabrication has been carried out.

For high-silica-content fibres, the fabrication method which produced fibres meeting all the requirements is based on a chemical-vapour-deposition process. This technique allows *in-situ* purification of the glass constituent materials, thereby permitting high-purity glass to be made without contamination. The success of these fibres has decreased the incentive for the development of other types of fibres from other glass systems, and superseded many fabrication techniques originally developed for making mixed-oxide glass fibres.

It is instructive to examine all the known techniques for fibre making in order to show the promises, problems and limitations.

6.1 From molten glass

Fibres can be made from molten glass by drawing upwards from a molten pool of glass, or downwards by letting the molten glass flow out of a nozzle at the bottom of a crucible. A clad fibre can be made by the same method if a cladding glass is allowed to flow over the core glass in an arrangement such as the concentric nozzles of a double crucible. (See Fig. 6.1).

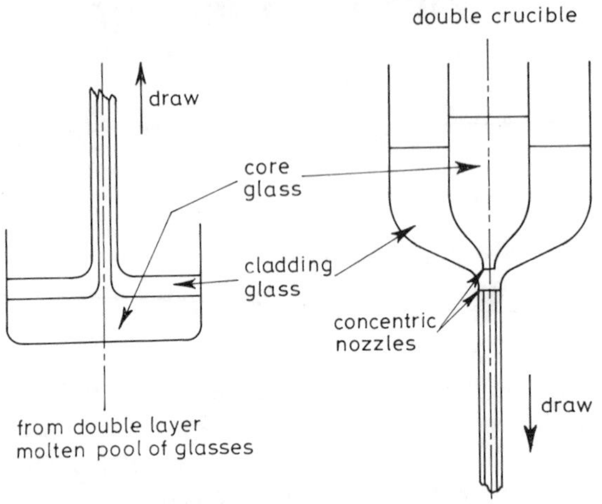

Fig. 6.1 *Making clad fibre from molten glass*

The principal prerequisites for producing low-loss fibre with acceptable geometry from molten glass are: (i) high-purity glasses, (ii) compatibility of core and cladding glasses in glass-transition temperatures and viscosity, and (iii) uniform fibre drawing rate. To meet the first two prerequisites, the first step is to obtain purified-glass-constituent materials. Then the material must be handled and mixed without introducing contamination before the glass-forming process, which involves melting the constituent materials at a high temperature. Here contaminants could be introduced from insulating material, heater, stirrer and the crucible. Furthermore, during this entire process, contamination can occur, when no purification action is present. The problem becomes acute if the glass melting temperature approaches the softening point of fused silica. In fact, above 800 °C, refractory materials for insulation become less plentiful and the vapour pressures of contaminants increase. Only a few heavy metals such as tungsten can serve as heater elements. RF induction heating is usually needed (since in RF heating the energy is coupled into near molten glass directly, and is therefore clean). Furthermore, the only materials suitable for crucibles are fused quartz and platinum. Compound glasses (mixed oxide glasses), such as borosilicate glass with relatively low melting points, can be made with reasonable purity.

After making the glass in bulk, homogenized glass rods can be produced by casting, i.e. pouring the glass into moulds. These rods are then remelted in a double crucible to produce fibres. During this process the glass may be contaminated by the impurities from the crucible. For some glasses electrolytic action could arise at the crucible nozzle, producing platinum particle inclusions in the fibre. One advantage of this process is the quality of the cladding–core interface. The liquid-to-liquid contact produces a perfect inter-surface. Even though diffusion of high-mobility ions will take place, this only means that the interface is graded in refractive index. Another advantage is the ability to produce fibre continuously in large quantities. The best fibre produced has a loss of about 5 dB/km at 0·85 μm.

6.2 From rod and tube

If the core glass is made in the form of a rod and the cladding glass in the form of a tube, a fibre can be made by placing the rod within the tube and collapsing the tube over the rod during or before fibre drawing. This method is an alternative to the double-crucible process, and has the same problems since pure glass must be made first. It has an added disadvantage in having a core–cladding interface which records the imperfections of the core rod surface and the inner surface of the tube. It was nevertheless an important technique for producing early fibres, and is still used as a technique to increase cladding thickness by sleeving a fibre preform with an extra cladding tube.

Other techniques for making fibre in compound glasses have been attempted. They are variations of the double crucible and rod-in-tube technique. Two examples are of limited significance. One is to make a rod with a glass containing a highly mobile ion as the dopant, such as thallium. The rod is then immersed in a hot salt bath containing a lighter mobile ion such as potassium. Thallium will now exchange with potassium to form a graded-index structure. This rod can then be drawn into a fibre with a graded-index core. The other method starts with a Vycor rod, an alkali-borosilicate glass containing a silica-rich network and an alkali and boron-rich region. The silica network will remain if the rod is immersed in a strong acid (except HF) while the other materials are removed. This silica network can be filled by a controlled high-purity material to form a graded-index fibre-core rod. Both techniques could be used to produce graded-index fibre with about 10–20 dB/km loss.

6.3 From chemical-vapour deposition (CVD)

High-silica fibre cannot be produced by the techniques discussed before. The temperatures involved at 1500—2000 ° are too high for crucibles. The eventual successful method was derived from the technique of chemical-vapour deposition of SiO_2 which had been used as a means to produce synthetic SiO_2 rod and SiO_2 film for coating semi-conductor devices. This technique involves glass forming at ~800 °C, fusing at 1500 °C and collapsing at 1900 °C, and drawing at 2100 °C.

It was found that vaporisable silicon compounds such as $SiCl_4$ can be decomposed in O_2 to form SiO_2 in a plasma (Fig. 6.2). This method was used as a means to deposit SiO_2 film as a passivation barrier over Si circuits. At the same time the fused-quartz makers developed techniques to make SiO_2 rods from $SiCl_4$ with a plasma torch or by reaction with an oxyhydrogen flame. The latter is called Verneuil hydrolysis. The silica produced contains up to 0·2% water. The starting material may be $SiCl_4$ or one of the silanes. The reaction is

$$SiCl_4 + 2H_2O \rightarrow SiO_2 + 4HCL \qquad (6.1)$$

The reaction within a plasma is an oxidation process involving no water:

$$SiCl_4 + O_2 \rightarrow SiO_2 + 2Cl_2 \qquad (6.2)$$

The reaction is highly exothermic and takes place at around 800 °C. The resultant SiO_2 in vitreous state deposits on a rotating pedestal and is fused by the plasma or the oxyhydrogen flame. SiO_2 grows as a cylindrical boule as the pedestal is slowly lowered. Large boules of tens of centimetres in length and several centimetres in diameter can readily be made.

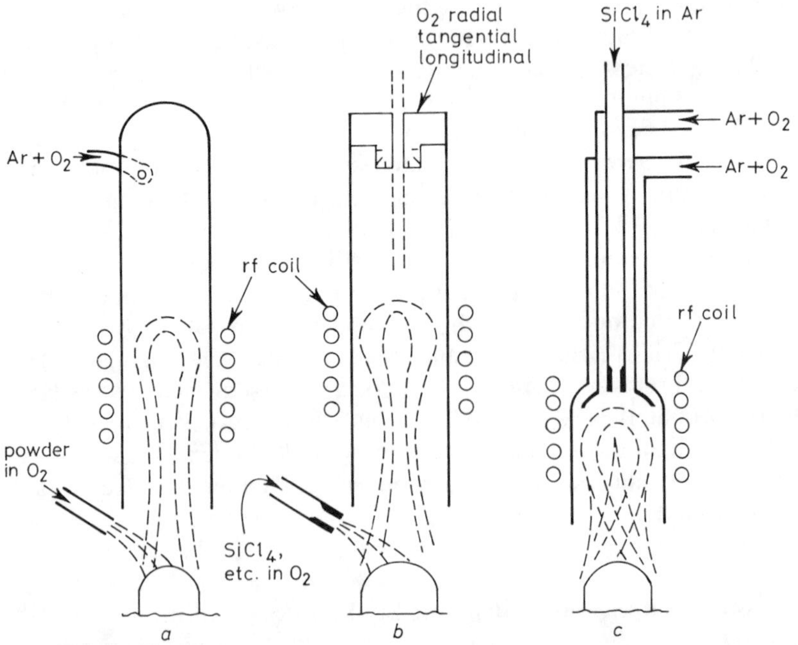

Fig. 6.2 *The Plasma torch as used for the preparation of vitreous silica or high-silica glasses: in its simplest form (a); as used by Nassau and Shiever (Ref. 1) (b); and as used by Kikuchi* et al. *(Ref. 2) (c)*

The CVD technique was adopted for producing low-loss fibre through a series of painstaking steps, eventually leading to three successful techniques. The most significant reason for this technique being suitable for fibre making is the occurrence of *in situ* purification during the process. This is due to the vapour-pressure difference between the glass-making materials and undesirable contaminating materials such as transition metals. Vapour pressures of metal halides are shown in Fig. 6.3.

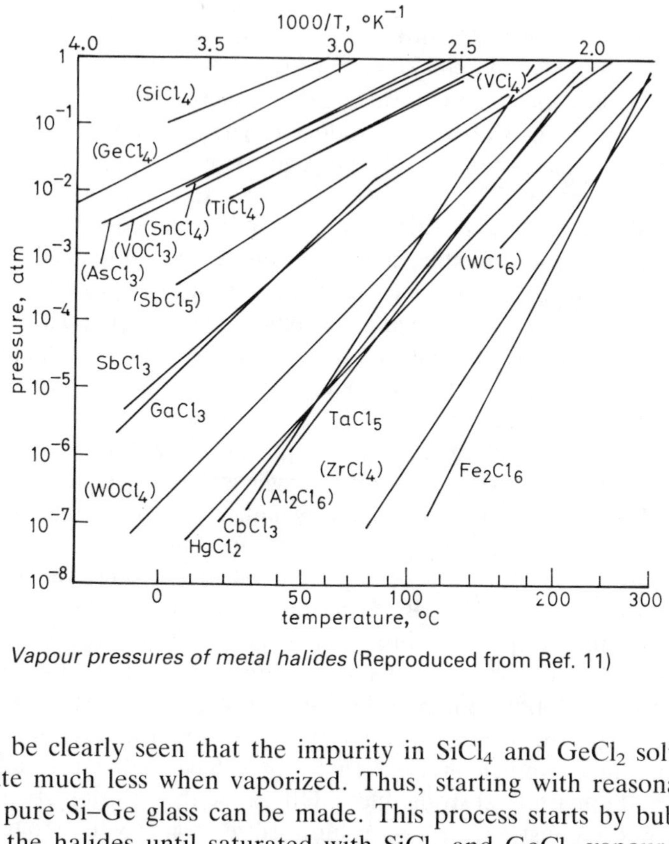

Fig. 6.3 *Vapour pressures of metal halides* (Reproduced from Ref. 11)

It can be clearly seen that the impurity in $SiCl_4$ and $GeCl_2$ solution will contribute much less when vaporized. Thus, starting with reasonably pure halides, pure Si–Ge glass can be made. This process starts by bubbling O_2 through the halides until saturated with $SiCl_4$ and $GeCl_4$ vapour. The gas mixtures in correct proportion are then reacted to form a SiO_2-rich glass with a controlled amount of GeO_2. The glass is in the form of a soot-like powder. This process has been developed into three highly successful forms. One involves the reaction of the glass-constituent within a SiO_2 tube. A second involves depositing the glass-constituent material on a mandrel. The third simply collects the material on a rotating and withdrawing platform to form a boule. All three forms of fibre making produce initially a fibre preform which is later drawn into the fibre with the desired geometry. Each calls for different forms of process control and processing steps. The difference in capital, space, energy and production rate are significant, but on aggregate all three are capable of producing cost-competitive high-quality fibres.

6.4 Modified chemical-vapour deposition process

The process in which the oxidation of halides takes place inside a SiO_2 tube is sometimes called 'inside process' or 'modified chemical-vapour-deposition process' (MOCVD).

In common with all CVD processes, the glass-constituent materials are liquid halides. $SiCl_4$ is the main glass-constituent material. Phosphorus in the form of $POCl_3$ and boron in the form of BBr_3 are sometimes used. Fluorine, in the form of Freon, can also be introduced. These materials are stored in all-glass or all-stainless-steel vessels. Oxygen gas is bubbled through each of the halides separately. The saturated vapours are channelled into a feed pipeline for delivery to the reaction chamber. When stainless-steel vessels are used, all traces of water must be removed from the constituents entering the system. Mass-flow controllers are used to regulate the exact amount of material to be delivered from each glass-constituent material. The mass-flow controllers measure the quantity of material delivered rather than merely measuring the flow rate of the oxygen carrier gas. Since saturation of the oxygen by the constituent material is of paramount importance, the way oxygen is bubbled through the halides is important. The storage vessel usually has significant levels of halide, while the oxygen is fed from near the bottom of the vessel. The vessel is to be kept in a constant-temperature region, and care is taken to ensure that the exit pipes do not become surfaces for condensation. The total flow rate is controlled by an additional oxygen line. The addition of He gas improves heat conductivity. It can be used to increase the uniformity of the reaction temperature. The entire material delivery system is programmed to deliver the required pre-mixed glass-constituent material in gaseous form. It is coupled to the reaction zone. For the inside processes the reaction zone is the inside of a quartz tube heated externally by a flame or internally with a plasma.

The quartz tube, when heated externally, must have uniform cross-sectional geometry and longitudinal straightness. Otherwise the reaction-zone temperature may not be constant, and the resulting fibre geometry, due to this and to tube cross-sectional variation, will not adequately meet practical requirements. The tube is mounted on a lathe bed and rotated. The reactants are introduced via a rotatable sealed joint. Fig. 6.5 is a schematic of this process.

The tube is heated by a controlled oxyhydrogen flame to a constant temperature up to 1800 °C. The flame traverses slowly from the gas input and towards the exit end. In the hot zone within the tube the halides oxidise to form SiO_2 and GeO_2. The composition of the resultant glass depends on the composition of the constituent materials and the operating temperature. The reaction is homogeneous and takes place in the gaseous phase. The resultant material clusters in the form of a soot and is carried downstream past the hot zone. Thermophoretic forces cause the soot to be deposited onto the inner surface of the tube. A percentage of the soot escapes into the exhaust. As the flame traverses over the deposited material and as it moves away from the input end towards the exhaust, it vitrifies the deposited soot into a layer of glass. Repeated traverses cause glass to form layer upon layer.

For making a step-index fibre, the deposited materials will have two constant compositions for the core or cladding region. First, the cladding material is deposited, and then the core material. Finally, the tube with the deposited material is heated until the surface-tension forces of the semi molten tube collapse the tube. The deposited material now becomes the core and cladding of a fibre preform supported within the quartz tube (Fig. 6.6).

The parameters affecting deposition rates are the flow of the reactant, the reaction time, the particle-size growth rate, thermal gradient, temperature uniformity, the stability of the heat source, and good dimensional consistency, precise alignment and straightness of the tube. In general, the reaction rate increases with temperature. For SiO_2, the reaction rate increases by an order of magnitude from 1200 °C to 1300 °C. At 1300 °C the formation rate can be assumed to be instantaneous. Particle growth rate is dependent on a coagulation constant of 10^{-2} seconds, which is required for particles to grow in size from 1 nm to 0·1 μm. The deposition is controlled by thermophoretic forces, which are dependent on the thermal gradient.

6.5 Thermophoretic deposition

Forces due to temperature gradient are caused by the molecular velocity imbalance. The radial velocity V_T is given by

$$V_T = -K(\nu V/T)\nabla T \qquad (6.3)$$

where

V = velocity of gas flow

K = thermophoretic constant

∇T = gradient of temperature

ν = viscosity

∇T = absolute temperature

K is dependent on the mean free path and particle size

For small particles formed during the MCVD process, $K = 0{\cdot}9$.

Assuming, uniform temperature T_h, rapid reaction completion and constant wall temperature downstream, T_{min}, the velocity profile is laminar. Hence

$$U(r) = V_{max}\,(1 - (r/R)^2) \qquad (6.4)$$

and

$$V_{max} = 2V_{av} = 2(Q/\pi R^2) \qquad (6.5)$$

where

R = tube radius

Q = total volume of gas flow

Thermal energy transport is given by

$$(1 - r^2)\,\frac{d\theta}{dz} = \frac{1}{P_e}\,\frac{1}{r}\,\frac{d}{dr}\left(r\frac{d\theta}{dr}\right) \qquad (6.6)$$

with

$$\theta = \frac{T - T_{min}}{T_h - T_{min}}$$

$\theta = 1$ at $z = 0$ and $\theta = 0$ at $r = 1$ and $z > 0$

r and z are normalised to R such that $r = 1$ at $r = R$

P_e = Pecket number = $V_{max}\,R/\alpha$

with

$$\alpha = \frac{k}{\rho c_p} \text{ thermal diffusivity}$$

k = thermal conductivity

ρ = density

c_p = heat capacity

The particles will have an axial velocity equal to the gas velocity and a radial velocity due to thermophoresis. From eqn. 6.3 and parameters derived from eqns. 6.4—6.6,

$$V_T = KP_r/(\theta + \theta^*)\left(\frac{d\theta}{dr}\right) \qquad (6.7)$$

where

$\theta^* = T_{min}/(T_h - T_{min})$

The Prendtl number $P_r = 0{\cdot}7$, K = 0·9
Deposition fraction is a function of KP_r and θ^*, and θ^* is more sensitive to T_{min} than T_h
The length of the deposition zone is dependent on total gas-flow rate and thermal diffusivity.

$$L \sim 0{\cdot}45\,P_e\,R \sim 0{\cdot}29\,(Q/\alpha)$$

This length is reached when gas and wall are at thermal equilibrium (see Fig. 6.4).

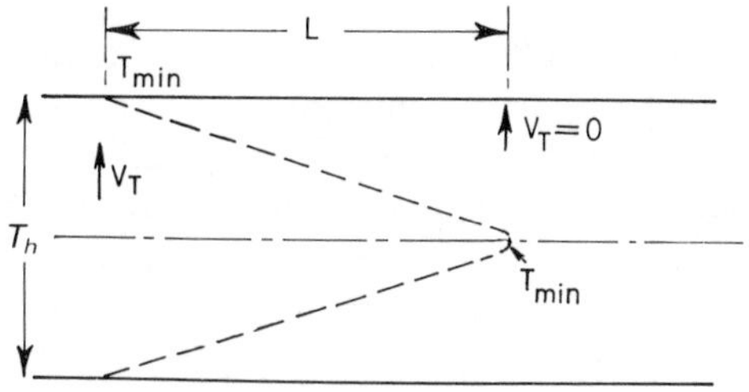

Fig. 6.4 *Gas and wall at thermal equilibrium*
For MCVD: T_h = 1750 °K, T_{min} = 450 °K

The deposition fraction of 0·5—0·65 has been measured, was 0·55 predicted by this thermophoretic theory.

It is advantageous to operate with a high-temperature hot zone in order to increase deposition rate. However, the highest operational temperature is governed by two factors: temperature at which the tube will collapse and secondly when excessive dopant evaporations are involved. In the first case the tube can be pressurised so that the temperature for the tube to collapse can be raised. In the second case GeO_2, the principal dopant which happens to be volatile is lost through the evaporation process. The depletion of Ge will take place when the deposited material is heated as the flame traverses the deposited material. This can be partially prevented by maintaining the high vapour presence of GeO_2 within the tube, but, in practice, this sets a ceiling to the temperature to which the tube can be heated.

The uniformity of the fibre depends on the constancy of heating. As deposition advances, the build up of material will influence the actual reaction-zone temperature. Hence, besides keeping the tube temperature constant through a feedback control loop, it is also necessary to increase the temperature as deposition proceeds. Furthermore, the tube must be mounted without residue strain in between well aligned chucks. Otherwise, the tube tends to warp after several deposition passes, thereby making temperature control more difficult.

A deposition rate of > 1 g/min can be routinely achieved with oxyhydrogen flame heating. The alternative to an outside flame is to use a plasma internal to the tube as the heating source. The temperature gradient in this case is much greater, leading to a higher deposition rate. Both RF plasma at normal pressure and microwave plasma at much reduced pressure have been used successfully inside a quartz tube for efficient chemical-vapour deposition. The RF plasma in a large-diameter quartz tube can achieve a 1—10 g/min deposition rate for pure SiO_2. The microwave plasma can achieve 100% efficiency deposition on a heated quartz tube at a relatively low temperature of less than 1000 °C. The chemistry involved is not perfectly understood. The deposited material is in fused form immediately. Owing to the lower temperature involved, the substrate tube need not be rotating and will not distort.

The inside deposition process suffers from the possibility of stress-induced fracture after the deposition has been completed but before the tube has been collapsed to form a rod preform. The deposited glass layer is under very high stress owing to the thermal-expansion-coefficient difference between the deposited material and the substrate material. The higher coefficient of the deposited material causes it to be in tension. Moving the tube before collapse can cause tube shattering. Hence the collapsing step should be performed immediately after the deposition step has been completed and on the same lathe.

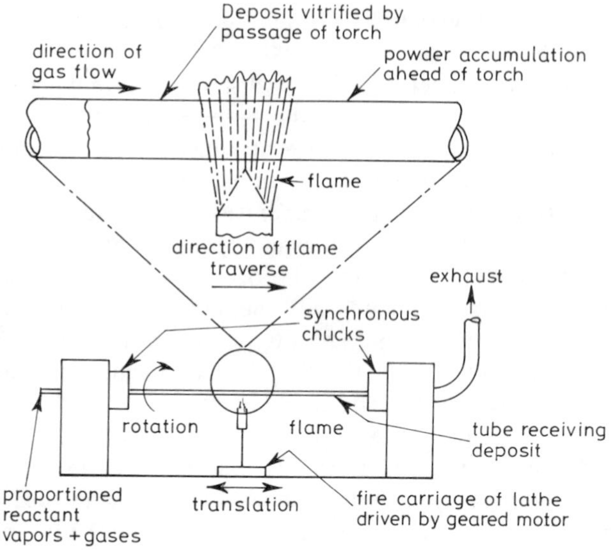

Fig. 6.5 *Schematic of the MCVD process.* (Reproduced from Ref. 11 of Chapter 4)

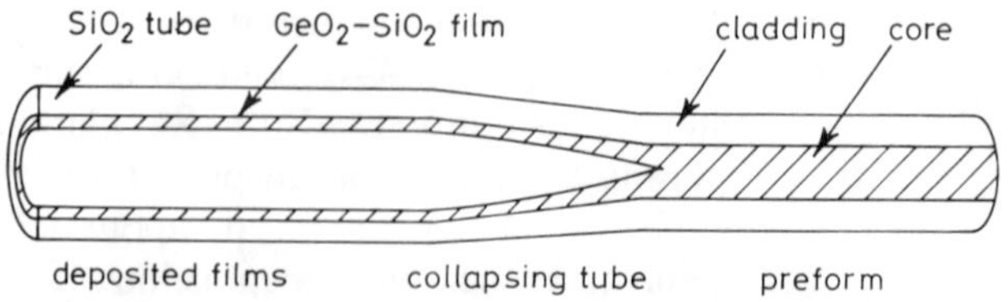

Fig. 6.6 *In the vapour-deposition method, the high-refractive-index core is deposited inside a silica tube. This material becomes the core after the tube is collapsed.* (Reproduced from Ref. 11 of Chapter 4)

6.6 Outside vapour-deposition process

The outside process involves the deposition of soot particles over a mandrel of quartz or graphite material. The glass-constituent materials in vapour form are dispensed from the equipment in a similar way to that used for the MCVD or inside process. The difference is how heat is applied. In the outside process, the glass-constituent materials are fed through a tube concentric to an oxyhydrogen coaxial torch. The torch design allows the reagents to be fed to the hot zone, where oxidation takes place and soot of desirable particle size is produced. The soot is sprayed on the mandrel layer after layer until a soot preform is made. The mandrel is then withdrawn. The soot preform is dehydrated, sintered and collapsed into a preform rod. The torch is also the source of heat to cause the soot to stick together.

Dehydration is necessary since this process of SiO_2 formation is a hydrolysis process. As the soot preform is porous, the dehydration can be very effective by heating the soot preform in Cl_2 gas. The chlorine gas combines with water to form HCl, which is carried away in vapour form. Sintering can then be done by raising the temperature further. During sintering the particles consolidate into a clear glass. Soot particle sizes of several hundred angstroms tend to pack into a porous solid with a suitable density for bubble-free sintering. The density is such that, during sintering, linear dimensions shrink by a factor of about 2.

A distinct advantage of this technique is in its ability to be scaled up in size easily. By using multiple torches and a larger mandrel, very high deposition rates of 10—20 g/min can be achieved and large preforms made. The process must be carefully controlled. The residual stress in the material must be evenly distributed so that successful sintering can be completed without the preform rupturing.

A mandrel-less outside deposition process is referred to as the vapour-axial-deposition (VAD) process. In this arrangement the deposition is on a withdrawing platform. The deposited material forms a lengthening boule. By arranging the torch position in relation to the boule centre, a graded-index-fibre core can be made. The soot boule is dehydrated and sintered as in the previous case. One difference is in having no mandrel to be withdrawn. Another difference is in the control of composition of the material deposited. The thermal profile of the soot boule and the flow of the incoming soot control the composition of the deposited material, even though the incoming soot, formed by a homogeneous reaction, would have a uniform composition. The advantage of the VAD technique is the possibility of making long preforms that can even be in a continuous form if dehydration and sintering can be done in tandem.

The resultant preforms produced by the outside deposition processes can be a complete composite fibre preform or a core rod with a relatively thin cladding. By sleeving the core with a quartz tube, a much larger preform is achieved. This application of sleeving is really a resurrection of the old rod-in-tube fibre-making technique. The techniques for high-silica-fibre fabrication all result in the formation of a preform which must be reduced to a suitable size for final fibre drawing. This can be done by re-drawing of the preform. Tailoring of a final preform dimension to the required final fibre OD and core diameter can be performed at this stage.

6.7 Fibre drawing

The preforms are drawn into a fibre by a fibre-pulling process. A fibre-drawing apparatus consists of a furnace, rod-feed mechanism, and fibre-pulling mechanism. The feed rate and pulling rate are fixed by the draw-down ratio. The heating zone of the furnace is profiled to provide a neck-down region suitable to balance surface-tension forces and viscous forces until uniform size reduction is achieved without instabilities (Fig. 6.8). A zirconium induction furnace is shown in Fig. 6.7.

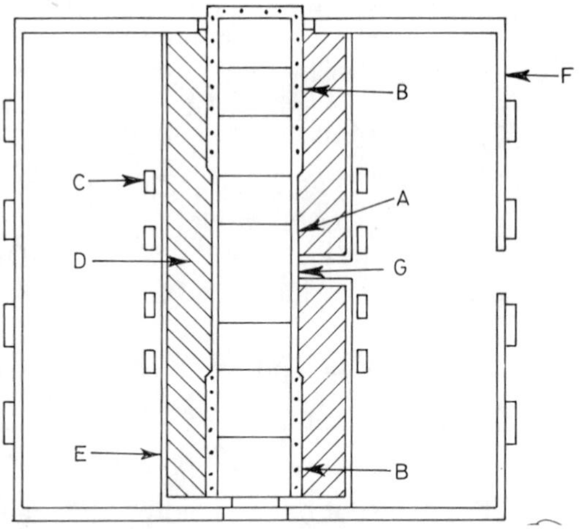

Fig. 6.7 *Vertical section of zirconia induction furnace*
A Zirconia susceptor rings
B Zirconia support rings
C Induction coil
D Zirconia insulation
E Fused quartz
F Copper shell
G Temperature monitoring
(Reproduced from Ref. 11 of Chapter 4)

A direct-heating graphite furnace is also available. The stability of fibre drawing depends on the mechanical stability of the mechanical system as well as the gas flow, particularly in the neck-down region where viscosity can change rapidly. A steady gas flow in the hot zone can prevent furnace-emitted

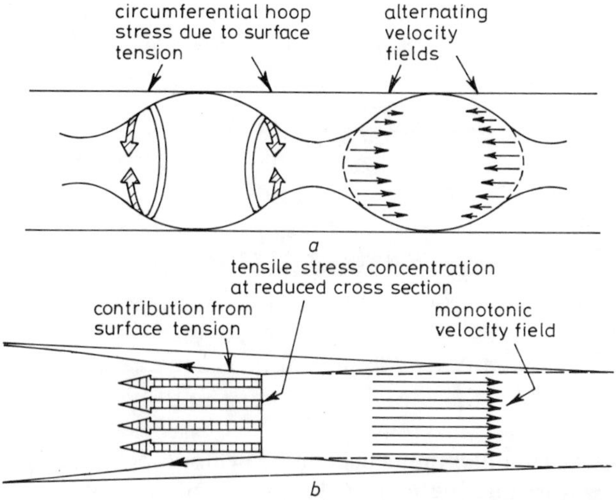

Fig. 6.8 *Comparison of (a) capillary and (b) tensile instabilities* (Reproduced from Ref. 3)

particles from hitting the fibre surface, thereby causing mechanical damage, and it can also ensure good dimensional stability.

In a fibre-drawing equipment, the freshly drawn fibre is uniformly protected by a dielectric coating. The coating material offers mechanical protection. The characteristics of the material are designed to be readily cured without appreciable changes in dimension, with good abrasion resistance, with the desired hardness, and to retain flexibility over a wide temperature range.

The coater design for high-velocity coating must be properly designed to avoid cavitation and turbulence. Coating at over 10 m/s can be achieved with some coating materials. In general, the fibre tends to self-center. The restoration forces, however, are small so that mechanical alignment of preform and coater nozzle is essential and must be controlled.

For high-speed draw the fibre must be force-cooled so that it can enter the coater at a low enough temperature. On-line fibre-diameter and concentricity measurements are needed. An effective method is based on illuminating the fibre transversely and observing the refracted and diffracted rays. The technique allows a feedback control loop to be established, which allows coating concentricity and fibre OD to be kept within a tight tolerance of 1—3%.

6.8 References

1 NASSAU, K., and SHIEVER, J. W.: 'Plasma torch preparation of high purity, low OH content fused silica', *Bull. Am. Ceramic Soc.*, 1975, **54,** p. 1004

2 KIKUCHI, B., *et al.*: 'Preparation of pure silica bulk for optical fibre', *Fujitsu Sc. Tech. J.*, 1975, **11,** p. 99

3 GEYLING, F. T.: 'Basic fluid-dynamic considerations in the drawing of optical fibres', *Bell Syst. Tech. J.*, 1976, **55,** pp. 1011–1056

Chapter 7

Fibre measurements

Measurement techniques are needed to allow fibre specifications to be verified both in the factory and in the field. From the fibre manufacturer's viewpoint, all fibres produced for delivery to the customers must meet the product specifications. When fibre is in an operational environment, the customer expects this specification to be verifiable. Since fibre performance is closely linked to light-source characteristics and launching conditions, the performance parameters are usually given for defined operational conditions. This applies particularly to multimode fibers. Measurement standardisation is required to bring together the manufacturer's specification and users' acceptance criteria. Spectral loss, dispersion, outside diameter, equivalent core diameters, numerical aperture, splice loss, and in addition, for single-mode fibres, effective cut-off wavelength for the higher-order mode, are the principal performance parameters of the fibres. Discussion of measurement techniques is best done for multimode and single-mode fibres separately. The similarities and differences can then be properly emphasised.

The major problem with multimode-fibre measurements is the dependence of the performance on the modal distribution within the fibre. After much effort a reasonably consistent way of measuring multimode fibre performances has been developed based on launching into the fibre a set of modes in an equilibrium distribution. This argues that multimode fibre always contains mode-conversion mechanisms and has differential mode loss owing to coupling to radiation modes, which accentuates loss of higher-order modes. Hence, after propagating over a long length, the modes in the fibre tend to settle down to an equilibrium set which is invariant in subsequent propagation. By launching the modes in equilibrium distribution, the loss and dispersion values should be more consistent. This also means that the NA for a multimode is not easy to measure since its effective value varies with length.

Theoretically, every mode in a multimode waveguide should have its own propagation characteristics, and can therefore be independently determined. If the coupling between modes is defined, the performance against distance is fully predictable. The argument of using an equilibrium mode distribution is justified when the coupling is a random process and the long-length characteristics of the fibre are to be determined. In practical situations, the equilibrium mode distribution is attained, but the transient region, where mode distribution departs from the equilibrium state, can be several kilometres long, depending on the perfection of the fibre, uniformity of external pressure on the fibre and the launching condition. With this as a caveat, the use of equilibrium mode distribution as a means to correlate performance between fibres is justified, even if the results represent a non-absolute characterisation.

The physical parameters specified by the manufacturer of fibres are: outside diameter, core effective diameter, core/outside diameter concentricity, ellipticity, jacket thickness and concentricity, and proof stress. The performance characteristics specified for graded-index multimode fibre are: spectral loss and dispersion, numerical aperture and maximum refractive index. For single-mode fibre these are: spectral loss and dispersion in single-mode operation, effective cut-off wavelength of the first higher-order mode, and mode volume. Conformation of the characteristics expressed by these parameters permit fibres made by different manufacturers to be used interchangeably. They also permit fibre performance in cable and in the field to be predictable. Thus, fibre cables and systems can be delivered to the users with a performance assurance. The performance parameters specified by the fibre cablemakers are the spectral loss and dispersion in the fibre cable, mechanical characteristics of the cable and service guarantees. The performance parameters specified by the system makers are usually in terms of the end-to-end system performance.

Discussion of fibre measurements will be confined to fibres only, and will not include process-monitoring techniques used by manufacturers unless otherwise relevant. It will also not include in any detail physical-parameter measurement techniques. The discussion will concentrate on the underlying principles of performance parameter measurements.

7.1 Relevant process-monitoring measurement

During fibre drawing, the fibre OD is monitored together with jacket concentricity as part of the manufacturing process. Thus records are available showing that along the entire fibre the OD and jacket OD and concentricity are met. After manufacturing, the fibre OD and jacket thickness and concentricity are checked by off-line optical-measurement techniques. A prestress test, either during or after fibre draw, is conducted to ensure that the fibre has passed the proof test. The uniformity of these parameters along the entire fibre length is important so that fibres can be packaged into cable and can be spliced with predictable performance if the fibre OD is used as the reference dimension for splicing. The proof test permits the service life of the fibre to be estimated.

7.2 Performance parameter measurements

7.2.1 Numerical aperture

The numerical aperture of fibre is defined as the sine of the acceptance angle, and is given by

$$\mathrm{NA} = \sqrt{n_1{}^2 - n_{cl}{}^2} = n_1\sqrt{2\Delta}$$

(see Chapter 2). For fibres with graded-index core, n_1 is interpreted as the maximum index n_{co} of the core and n_{cl} is the cladding index. For multimode graded-index fibre, it is sometimes convenient to have the concept of a local NA at r expressed as

$$NA(r) = (n^2(r) - n_{cl}{}^2)^{\frac{1}{2}}$$

For single-mode waveguide, NA influences loss, core size for a chosen operating V value, and bending sensitivity. Also, it influences the mode volume or mode field diameter. A design for a given mode volume will fix the actual NA for a given core-index profile, for which an equivalent step-index core can be defined. The measurement of mode field diameter will be discussed later. Hence, for single-mode fibre NA is a secondary parameter when direct measurement is not meaningful.

For multimode waveguide, NA is much higher and is an important parameter. It influences loss, and more importantly, bend sensitivity. Measurement of NA is standardised, and is made by obtaining the near-field and far-field pattern of a fibre under specific launching conditions.

In the near-field measurement, the test fibre is typically 2 m in length and is fully excited by over-filling the fibre with a light beam whose spot size exceeds the fibre diameter, and whose divergence angle exceeds the fibre NA. The cladding modes are stripped by passing the fibre through an index-matching liquid or over a section painted with an absorbing material. The near field is projected through a lens and a detector, and a small pinhole aperture is used to scan and record the near-field intensity at the image plane. Since the near field is a function of the modal power distribution, the measurement of near-field intensity distribution can be converted to fibre NA. The apparatus is as shown in Fig. 7.1.

Since power acceptance at a point τ for the core centre is proportional to the local NA,

$$\frac{P(r)}{P(0)} = \frac{n^2(r) - n^2(a)}{n^2(0) - n^2(a)} \tag{7.1}$$

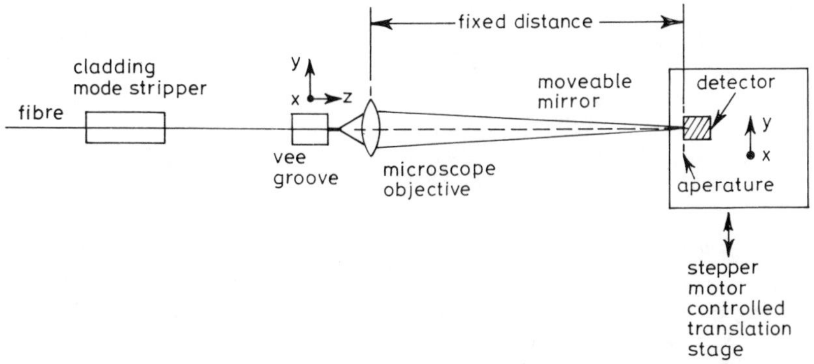

Fig. 7.1 *Near-field measurement system based on a radial scan of a magnified near-field image where the launching optics are not shown* (Reproduced from Ref. 1)

the $n(r)$ can be deduced. However, this requires that all circularly symmetric modes are equally excited. It also assumes that, over the short distance, no significant mode coupling has taken place and no differential mode attenuation has resulted. This situation is true if the excitation source is Lambertian and no leaky modes are excited. If leaky modes are present the results will be misleading. Furthermore, the finite number of modes will also give rise to ripples which could be confused with the actual ripple of the refractive-index profile. The ripple period can be shown to be proportional to $1/nc$.

This technique has many problems and is seldom used for NA measurement, but has been accepted as a standard method of measuring the core diameter by the Electronic Industries Association (EIA). In a modified form in which the refracted near field instead of the transmitted near field is measured, the fibre diameter and refractive index profile can be measured. The latter has been adopted by CCITT as the reference test method for diameter and profile measurement.

The refracted near field avoids the leaky-mode problem by measuring only the refracted radiative field. Light from a laser is focused onto the end of the fibre with angular distribution including angles much larger than the fibre acceptance angle. The guided modes are led away and the leaky modes are blocked from the detector by a disc. Only the refracted field is monitored by collecting with a lens system. The fibre end is scanned across the light launching spot. The refracted light is proportional to the local NA. The apparatus is as shown in Figs. 7.2 and 7.3, and a typical result is as shown in Figure 7.4.

Far-field pattern is a direct measurement of NA. The angular distribution is measured by a detector rotating at a distance D away from the fibre end with the fibre end as the centre of rotation, as shown in Fig. 7.5. A typical result is shown in Fig. 7.6.

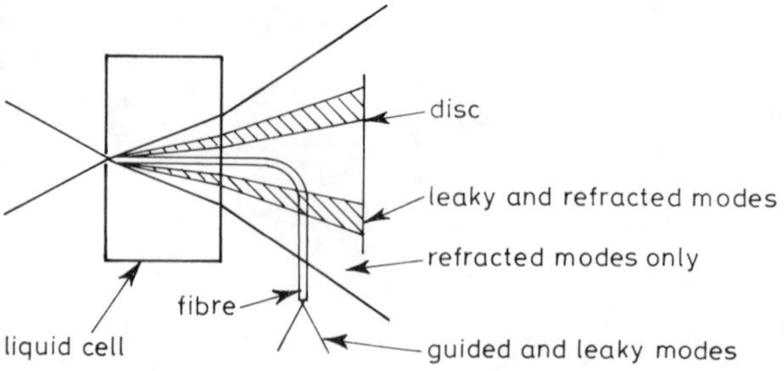

Fig. 7.2 *Refracted near-field technique: schematic diagram* (Reproduced from Ref. 2)

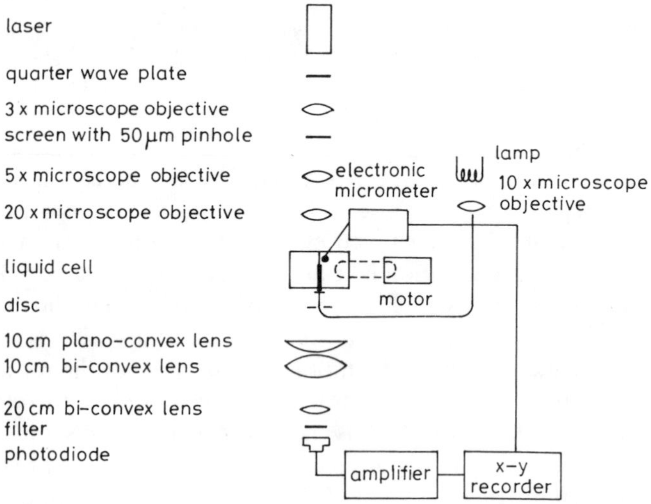

Fig. 7.3 *Refracted near-field technique: schematic diagram of apparatus* (Reproduced from Ref. 2)

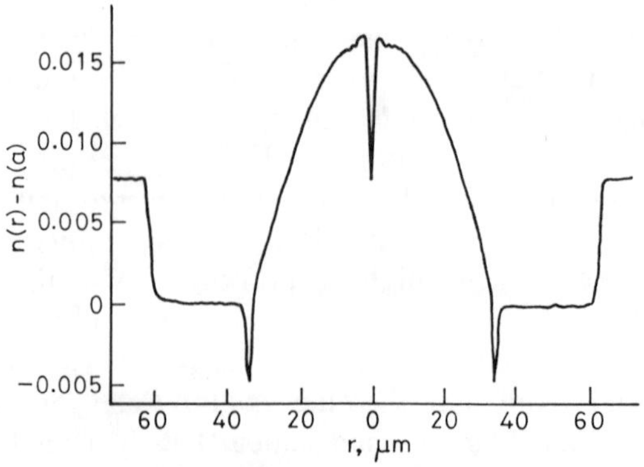

Fig. 7.4 *Index profile of a graded-index fibre obtained by the refracted near-field technique* (Reproduced from Ref. 2)

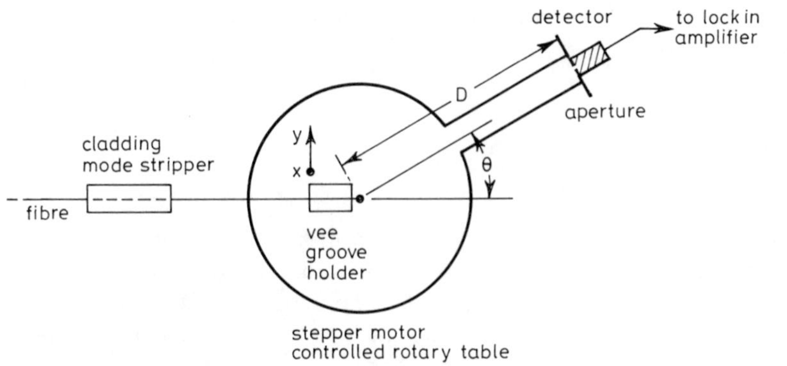

Fig. 7.5 *Far-field measurement system using a fixed fibre end and a rotating detector* (Reproduced from Ref. 12 of Chapter 4)

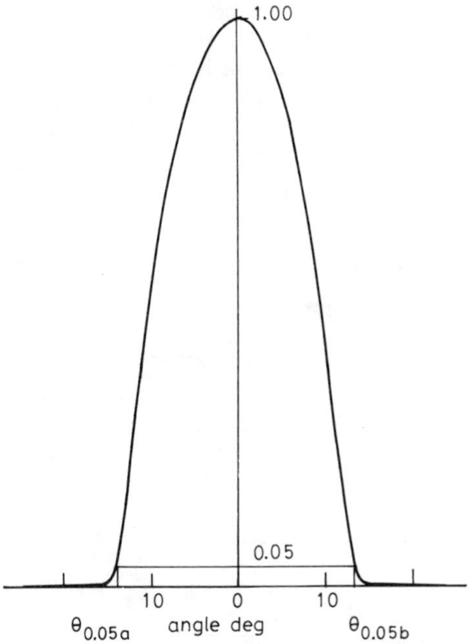

Fig. 7.6 *Normalised far field pattern for graded-index fibre* (Reproduced from Ref. 12 of Chapter 4)

This has been standardized by EIA as the reference test method and by CCITT as the alternative test method. A 2 m length fibre with full aperture excitation, as in the near-field case, is used. The NA is given as the angle between the power levels at 5% of the maximum value and the maximum value. The truncation at 5% results in a departure from the theoretical NA of about 10%, but avoids leaky-mode effects. The theoretical NA can be indirectly deduced from near-field measurement or from dopant-concentration measurement of the fibre-core material. Both need not be absolute measurements.

7.3 Loss measurement

7.3.1 Cut-back method

This is essentially an insertion-loss measurement in which the transmitted power of two lengths of fibre are compared. The input condition must be invariant for these two measurements. In practice, the fibre to be measured is attached to the spectral source via a launching fibre which delivers light with an equilibrium mode distribution to the fibre. This is made possible by scrambling the fibre modes through a mode-scrambling section consisting of a fibre subject to random bands (simulated by sandwiching the fibre between rough surfaces under pressure) or of a special section of fibres, typically a step-graded arrangement, which cause the output to be in near-equilibrium mode distribution. The fibre on test of length L is coupled to the launching fibre for a fusion or glued splice, as a butt joint is called. The transmitted power is measured over the spectral range as the wavelength of the source is scanned by a monochrometer or with the use of a series of discrete bandpass filters. Then the test fibres, except for a short length, is cut off, and the output from the short section is measured over the same spectral range (Figs. 7.7 and 7.8). The spectral attenuation is given by

$$A(\lambda) = 10 \log_{10} [P_L(\lambda)/P_z(\lambda)] \text{ decibels} \tag{7.2}$$

If z is small compared with L, the attenuation per unit length is A/L. The loss measured is the total loss including both the absorption loss and scattering loss. It is influenced by the launching condition and the mode coupling along the length and is averaged over the fibre length. This technique has been standardized by EIA and CCITT.

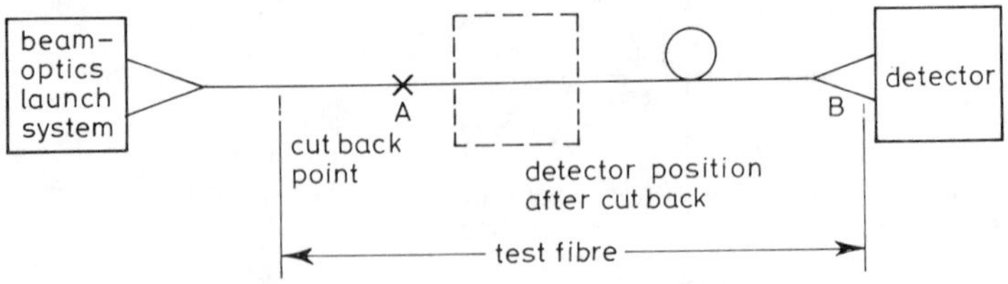

Fig. 7.7 *Attenuation measurement by cutback in the beam-optics launch case* (Reproduced from Ref. 12 of Chapter 4)

For single-mode fibre the mode-scrambling section is replaced by a single-mode-fibre section; otherwise the procedure is identical. A typical loss curve is as shown in Fig. 7.9.

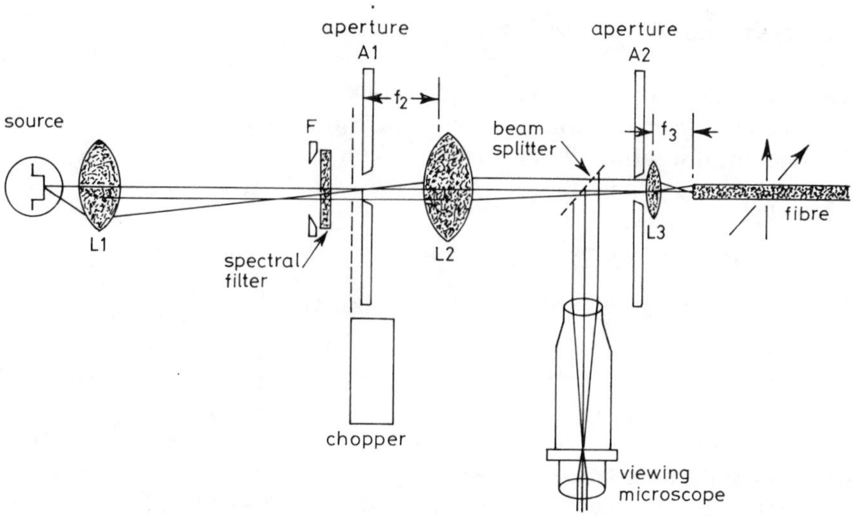

Fig. 7.8 *Light launching with the beam-optics method*
L_1, L_2 and L_3 are lenses
A_1 and A_2 are apertures
F is a spectral filter
F_2 and F_3 are the focal lengths of lenses L_2 and L_3, respectively
(Reproduced from Ref. 3)

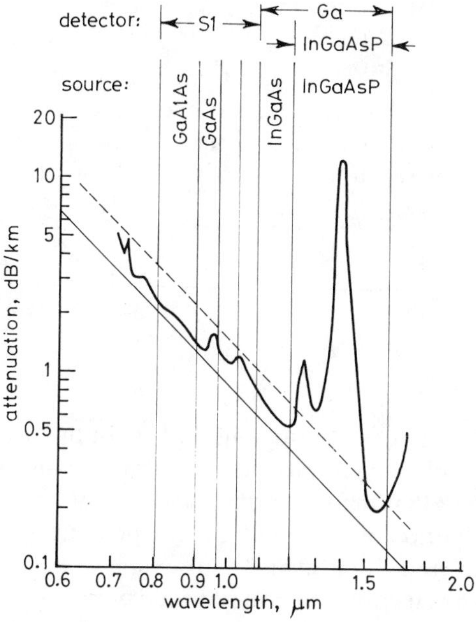

Fig. 7.9 *Optical fibre attenuation versus wavelength.* (Reproduced from Ref. 23 of Chapter 5)

7.4 Optical time-domain reflectometry

If an impulse is launched into one end of a fibre, the back-scatter energy of the pulse from the fibre contains the length-resolved information of fibre attenuation, discontinuities etc. Transmission loss can be derived as follows:

$$P(z) = P_i \exp\left[- \int_t^z \alpha_i(x)dx\right]$$
$$P_s(z) = R(z)\, P(z) \exp\left[- \int_0^t \alpha_s(x)dx\right]$$
$$= R(z)\, P_i \exp\left[- \int_0^z (\alpha_i(x) + \alpha_s(x))\, dx\right]$$

with

$$t = 2nz/c$$

where $R(z)$ is the reflection coefficient at z

The loss coefficient $(\alpha_i - \alpha_s)$ is obtained from

$$\frac{d}{dx}\left[\ln\left(\frac{P_i}{P_s}\right)\right] = \alpha_i(z) + \alpha_s(z) \qquad (7.3)$$

Therefore the slope gives the sectional loss. Typical OTDR data is as shown in Fig. 7.10. The equipment is as shown in Fig. 7.11.

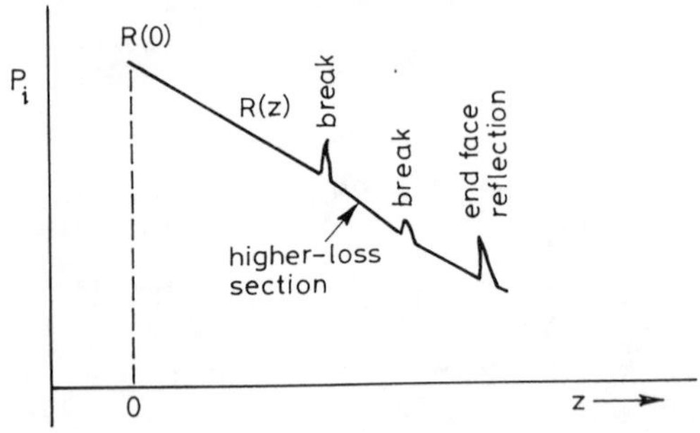

Fig. 7.10 *Typical OTDR data*

The cell can be a semi-transparent mirror. In that case the initial reflection from the fibre end may saturate the detection system. A gating switch may be needed to ensure that the saturation tail does not obscure the reflected signal over an excess period of time, and thereby prevent the initial part of the fibre from being analysed. By using a 4-port coupler this problem can be avoided. With a sensitive detection system a returned pulse 30—40 dB down due to back scatter can be identified. This is sufficient to analyse the repeaterless span of a fibre system.

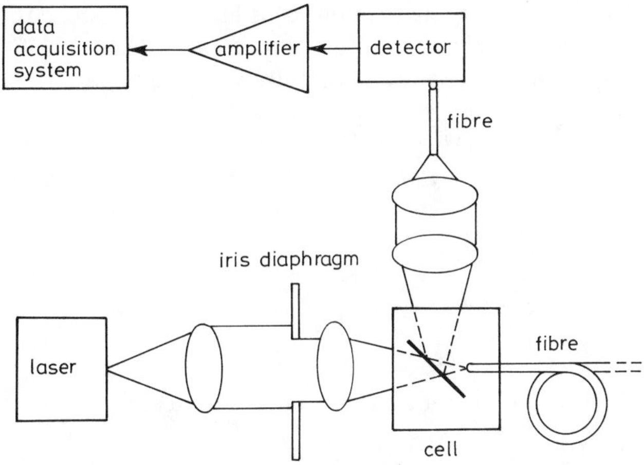

Fig. 7.11 *Experimental arrangement of back-scattering measurement apparatus.* (Reproduced from Ref. 4)

OTDR is a diagnostic method to analyse the loss of the entire fibre and to detect flaws including breaks. It is a non-destructive test method suitable also for field use. The loss measured is usually at a single wavelength and corresponds to the case of illuminating with a Lambertian source. Hence, for multimode fibre, the loss measured is higher than the value measured for an equilibrium mode distribution.

7.5 Dispersion measurement

The transmission bandwidth available for information transmission along a fibre is governed by the dispersion characteristics of the fibre. The bandwidth can be measured directly by a frequency-domain method or indirectly via the impulse response. For digital pulse propagation the impulse response directly measures the pulse broadening, and therefore gives the maximum permissible bit rate of transmission.

7.6 Time-domain measurement

If the spectral spread due to the modulation is small compared with the original spectral width of the light source (incoherence ensured), and if the spectrum is stable for the entire duration of the pulse, the power of the pulses within the fibre can be added linearly if overlap occurs. Under these conditions, the output pulse power P_{out} can be calculated through the convolution of the input pulse P_{in} with the power impulse response of the fibre $h(t)$. Therefore

$$P_{out} = P_{in}(t)^{*}h(t) \text{ or } = \int_{-\tau/2}^{\tau/2} P_{in}\,(t-\tau)\,h(\tau)d\,\tau \qquad (7.4)$$

Thus, if a narrow input pulse (impulse) is launched through the end of a fibre and the output pulse is detected, output pulse shapes define the impulse response function. Define an RMS pulse with σ given by:

$$\sigma^2 = \frac{\int_{-\infty}^{\infty} (t-\bar{t})^2 P_{out}(t)\, dt}{\int_{-\infty}^{\infty} P_{out}(t)\, dt} \tag{7.5}$$

and $\bar{t}$ expressed as

$$\bar{t} = \frac{\int_{-\infty}^{\infty} t P_{out}(t)\, dt}{\int_{-\infty}^{\infty} P_{out}(t)\, dt}$$

In general, this step must be carried out numerically.

If $P_{in}(t) = \exp - (2t/\sigma_{in})^2$

and $P_{out}(t) = \exp - [2(t - \bar{t})/\sigma_{out}]^2$ are Gaussian

then $h(t) = \exp - [2(t - \bar{t})/\sigma]^2$ is Gaussian

and $\sigma_{out}{}^2 = \sigma_{in}{}^2 + \sigma^2$

RMS addition is demonstrated. This result is generally consistent with the definition of σ. The construction of h_t from P_{in} and P_{out} is computed incrementally by

$$y(t) = 0 \quad 0 \leqslant t < t_0$$

$$\begin{aligned} y_1 = y(t_0 + \Delta t) &= \int_0^{t+\Delta t} h(t_0 + \Delta t - \tau)\, x(\tau)\, d\tau \\ &= \int_0^{\Delta t} h(t_0 + \Delta t - \tau)\, x(\tau)\, d\tau \\ &= h_1 x_1 \Delta t \end{aligned}$$

$$y_2 = y(t_0 + 2\Delta t) = (h_2 x_1 + h_1 x_2)\,\Delta t$$

$$\vdots$$

$$y_n = y(t_0 + n\Delta t) = \left(\sum_{k=1}^{n} h_{n-k+1}\, x_k\right)\Delta t$$

Or

$$Y = XH\Delta(t)$$

Therefore

$$H = \frac{1}{\Delta(t)} X^{-1} Y \tag{7.7}$$

To prevent the function H from diverging, we have a new matrix with x and y values taken to $l > n$, and take $h_k = 0$ for $k \geqslant n+1$. The resultant matrix equation has no exact solution. Using the least-squares method and taking $dP/dH = 0$, H_{opt} is derived:

$$P = [(Y/\Delta t) - XH]^t[(Y/\Delta t) - XH] \tag{7.8}$$

$$H_{opt} = (X^t X)^{-1} X^t Y\left(\frac{1}{\Delta t}\right) \tag{7.9}$$

The impulse response must be integrated to form the step-response function. Then, by differentiating the response, the impulse function can be computed.

The foregoing describes the general method for computing an impulse response. However, in the measurement of the transmission characteristics of an optical fibre, H_{opt} thus obtained consists of widely scattered points. The following smoothing technique (which is a standard technique in the field of system identification) should be employed to reduce scattering [Ref. 5].

Fig. 7.12 shows an example of the scattered impulse response computed using eqn. 7.9. Generally, by integrating an impulse response with respect to time, we obtain the step response, and by differentiating the latter, we obtain the former. If we integrate the data in Fig. 7.12 which looks hopeless, we obtain the step response shown in Fig. 7.13. Note that the ordinate of the step response shown in Fig. 7.13 is normalised to unity.

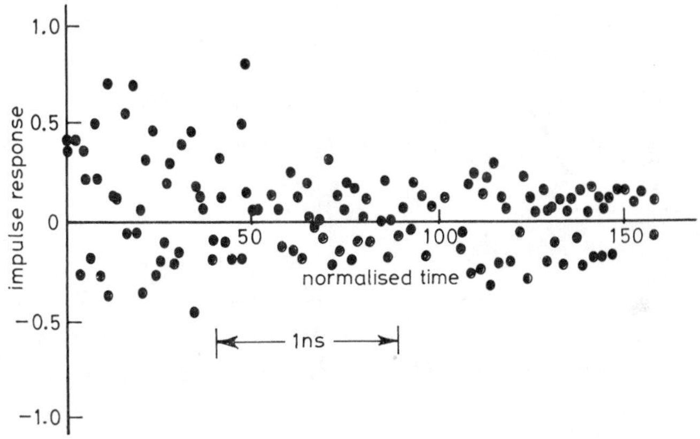

Fig. 7.12 *An example of the 'scattered' impulse response obtained by using eqn. 7.9.* (Reproduced from Ref. 7)

It is not difficult to smooth out scattering in the step response shown in Fig. 7.13; e.g. we can average the data locally to obtain a smoother response. The result of such smoothing is shown in Fig. 7.14. The impulse response can then be computed by differentiating Fig. 7.14; the result is shown in Fig. 7.15.

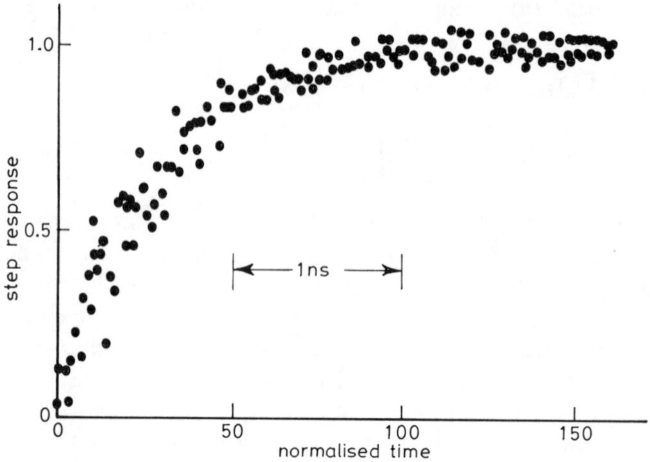

Fig. 7.13 *Scattered step response obtained by integrating the impulse response shown in Fig. 7.12.* (Reproduced from Ref. 7)

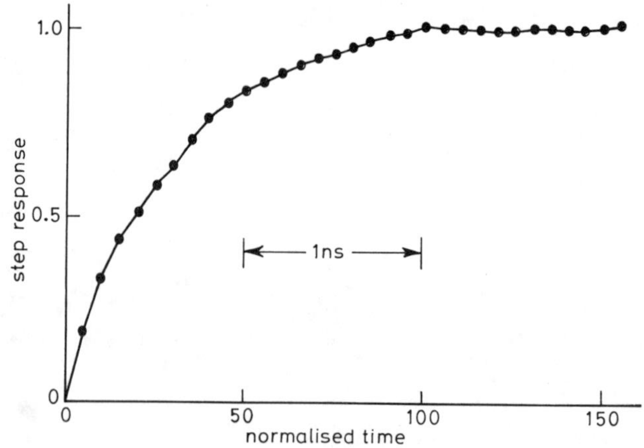

Fig. 7.14 *Step response obtained by smoothing Fig. 7.13.* (Reproduced from Ref. 7)

To prove the validity of the impulse response thus obtained, we synthesise the output waveform by convoluting the input waveform and the computed impulse response, and compare the result with the measured output waveform. Fig. 7.16 shows an example of such comparison; the solid and dashed curves show the measured and synthesised output waveforms respectively. The overall shapes of the waveforms show good agreement, but small ripples are missing in the synthesised waveform.

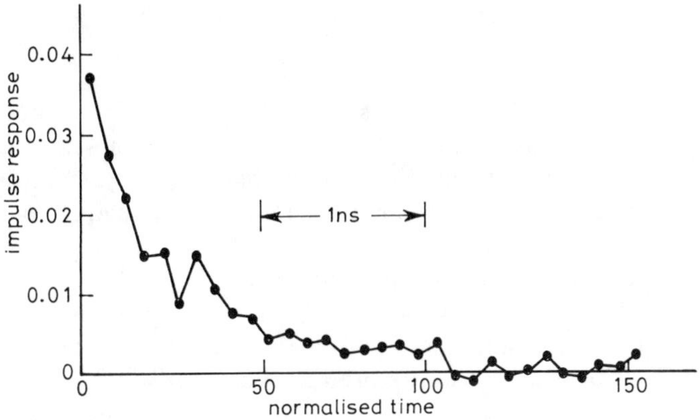

Fig. 7.15 *Impulse response obtained by differentiating the response of Fig. 7.14.* (Reproduced from Ref. 7)

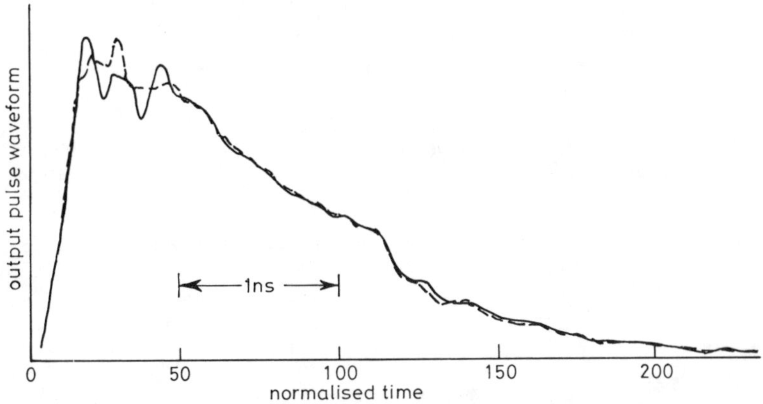

Fig. 7.16 *Measured output waveform (solid curve) and synthesised output waveform (dashed curve) obtained by convoluting the input waveform and the computed impulse response.* (Reproduced from Ref. 6)

7.7 Frequency-domain measurement

The Fourier transform of the impulse response

$$H(f) = \int_{-\infty}^{\infty} h(t)\, e^{-i\omega t}\, dt \tag{7.10}$$

and

$$h(t) = \frac{1}{2\pi}\int_{-\infty}^{\infty} H(f)\, e^{-\mathrm{iwt}}\, dt \tag{7.11}$$

$H(t)$ can be measured in the frequency domain in both amplitude and phase.

$$H(f) = |H(f)|\, e^{i\,\theta(f)} \tag{7.12}$$

A typical experimental set up is as shown in Fig. 7.17. The source is directly modulated or externally modulated. A vector voltmeter or a network analyser can be used to measure the amplitude and phase for modulation frequencies up to several gigahertz. The relative amplitude is readily measured for increasing modulation frequency up to, say, the 3 dB point. The phase, however, is extremely difficult to measure, since the total phase change of the long length of fibre is very large while the differential phase is very small. The frequency-domain measurement is often carried out for amplitude response only. The 3 dB bandwidth can be used as a comparison measurement (Fig. 7.18).

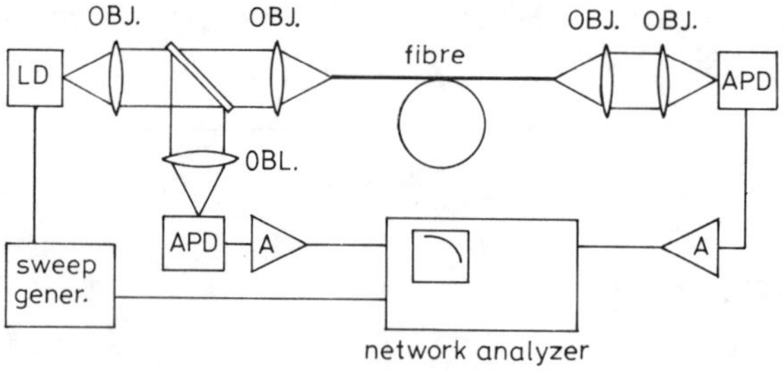

Fig. 7.17 *Schematic for direct measurement of the frequency transfer function.* (Reproduced from Ref. 8)

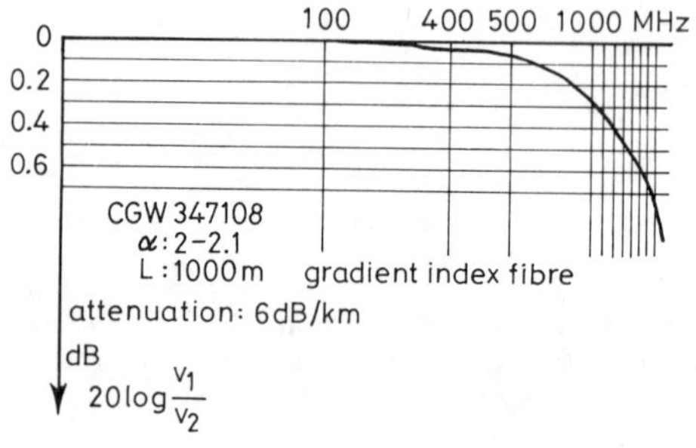

Fig. 7.18 *Amplitude of a graded-index fibre transfer function* (Reproduced from Ref. 9)

The impulse measurement can, of course, be converted to frequency response via the Fourier transform. But the amplitude frequency response cannot be converted to impulse response without the phase information. The most difficult issue with regard to the impulse response or frequency transfer function of a fibre is the length dependence. If a fibre is measured in 1 km lengths, the response of an n kilometre fibre made up of $n \times 1$ km lengths is unlikely to be predictable, especially if the fibres are graded-index fibres with index power α around the optimum value. The concatenation of a random selection of such fibres will result in total dispersions which varies from L^n with $0 < n < 2$.

For single-mode waveguide the dispersion due to the waveguide mode dispersion is small. The major contribution is due to material dispersion. A measurement technique is to use a Raman laser to generate a continuous spectrum of Stokes frequencies. A monochrometer is used to select the pulse wavelength. The delay through the fibre at each wavelength is measured. Dispersion is expressed in ps/km nm. The apparatus is as shown in Fig. 7.19, and a typical result is given in Fig. 7.20.

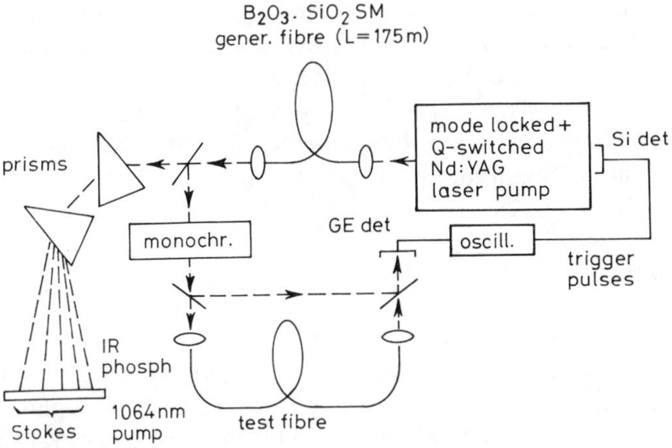

Fig. 7.19 *Experimental arrangement for measuring transmission-time delays versus wavelength using stimulated Raman scattering in a single-mode fibre* (Reproduced from Ref. 10)

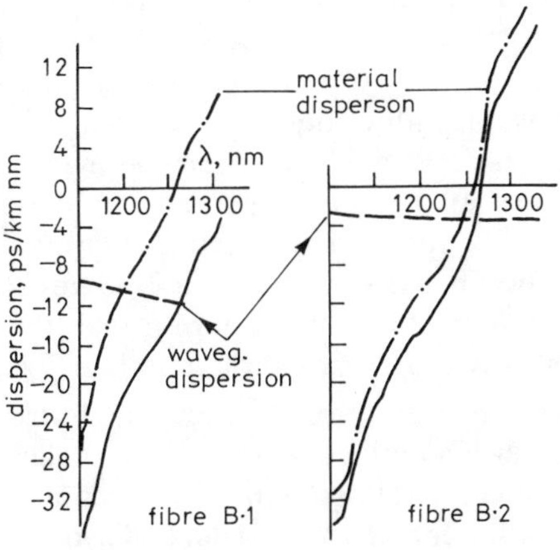

Fig. 7.20 *Single mode fibre dispersion (——) versus wavelength*
Waveguide dispersion (- - -) is subtracted to obtain material dispersion (–·–·–). Fibre B-1 causes more waveguide dispersion than fibre B-2 (Reproduced from Ref. 10)

7.8 Measurement of single-mode characteristics

For single-mode fibres two measurements which characterise the single-mode performance are important: cut-off wavelength and mode-field diameter. The cut-off wavelength of the first higher-order mode defines the wavelength above which the fibre operates in a single mode for a fibre with a perfect circular core. The mode-field diameter is a measure of the field extending from the centre of the core. Field extension at a given wavelength varies with the operating V value. It influences loss, bend sensitivity and splicing loss, particularly if the cladding material has a higher loss than the core material. For single-mode fibre with profiled core, non-circular core or birefringent core, the cut-off wavelength and mode-field diameter are still valid parameters for describing the fibre characteristics. However, the performance is specified for a reference-fibre-axis orientation only. Single-mode fibre is specified with a cut-off wavelength and a mode field diameter for a specific wavelength so that its property at an operating wavelength is fully characterized.

7.9 Effective cut-off wavelength

The theoretical cut-off wavelength of a single-mode fibre is usually not measured, since the stability of the lowest-order mode and the first higher-order mode near cut off of the latter is poor. Coupling of energy between the modes and the radiation field makes precision measurement over a length of fibre impractical. Thus, when the presence of the higher-order mode, which in a weakly guided-mode approximation is the LP_{11} mode, can just be reliably detected, the wavelength is already shorter than the theoretical cut-off wavelength. This wavelength is referred to as the effective cut-off wavelength. The precise definition is by specifying the power of LP_{11} mode relative to the LP_{01} power to be a given value for the equivalent cut-off wavelength for a given length of fibre with a given radius of curvature.

The measurement is performed on a nearly straight standard fibre of 2 m in length, with curvature not less than 14 cm radius of curvature. The fibre must be loosely wound on a spool. The spectral transmission is measured and re-measured after the fibre is wound over a 2—3 cm-diameter mandrel. The plot of the ratio of the two transmitted powers over the spectral region gives the effective cut-off wavelength as indicated in Fig. 7.21.

The curvature causes the cut-off to occur at a shorter wavelength. The longer-wavelength edge is defined as the effective cut-off wavelength at the point where the differential loss is 0·1 dB. This arrangement allows the repeatability to be about 2 nm. EIA and CCITT have adopted this method as a standard measurement.

A second method is based on the measurement of mode-field diameter (MFD) with wavelength. A spectral plot of the MFD indicates a sudden

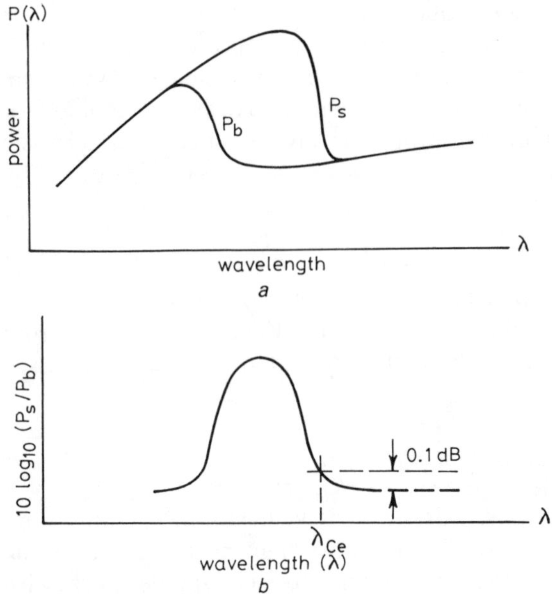

Fig. 7.21 *Single-bend attenuation technique*
a Power transmitted through bent (P_b) and straight (P_s) fibres vs. wavelength.
b Attenuation of bend and definition of effective cut-off wavelength λ_{ce}.
(Reproduced from Ref. 12 of Chapter 4)

increase of MFD as the LP_{11} mode is excited; and as wavelength continues to decrease the mode-field diameter first increases and then decreases. The effective cut-off is derived by taking the intersection of the two slopes adjacent to the minimum-diameter point.

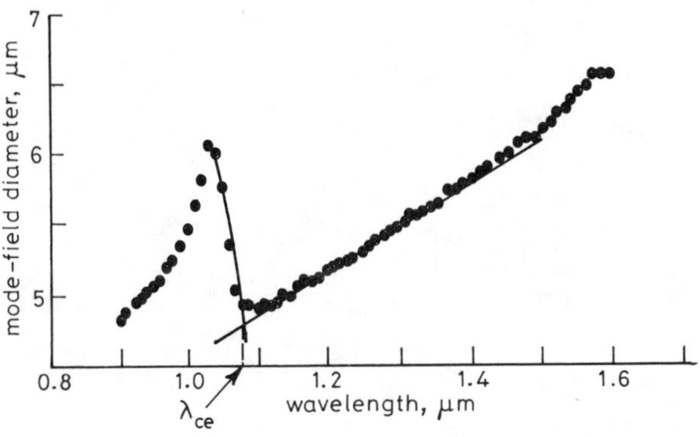

Fig 7.22 *Obtaining effective cut-off wavelength from spectral scan of mode-field diameter* (Reproduced frm Ref. 11)

The MFD technique can be modified to check quickly the cut-off wavelength. In one modification the fibre on test is butt-spliced to another single-mode fibre with much shorter wavelength. As the spectrum is tuned, near cut-off the splice losses vary rapidly. This avoids the time-consuming procedure of measuring accurately the spectral transmission over a broad spectral region.

7.10 Mode-field diameter

At a given wavelength, the fibre-core diameter, core/cladding relative refractive indices and the operating V value are fixed. This operating V value determines the mode-field distribution and hence its associated diameter. Since

$$V = \frac{2\pi a}{\lambda}\sqrt{n_{co}^2 - n_{cl}^2}$$

single-mode fibres with different values of a, n_{co} and n_{cl} may have equal V value and give equivalent performance. In fact, even for a fibre with a graded-index core, an equivalent single-mode fibre with an equivalent step-index core diameter and NA can be prescribed.

MFD is defined as the diameter of the optical field carried by the fibre. Three alternative definitions are possible. First, the width of MFD is defined in terms of the radius ω_b, which is the radius to the $2e^{-2}$ point of the optical power E^2 of a Gaussian field:

$$E_g(r) = E_0 \exp\left(-r^2/\omega_b^2\right)$$

ω_b is chosen so that the launched-power efficiency given by the integral I is a maximum:

$$I = \frac{\left[\int_0^\infty r\,E_g(r)\,E(r)\,dr\right]^2}{\left[\int_0^\infty r\,E_g^2(r)\,dr \int_0^\infty rE^2(r)\,dr\right]}$$

where

$E(r)$ is the actual fibre field distribution

The other two definitions are in terms of two other radii given by ω_j and ω_m:

$$2\omega_j = 2\left[\frac{2\int_0^\infty E^2(r)\,r\,dr}{\int_0^\infty \left(\frac{d\,E(r)}{d\,r}\right) r\,dr}\right]^{\frac{1}{2}}$$

and

$$2\omega_m = 2\left[\frac{2\int_0^\infty E^2(r)\,r^3\,dr}{\int_0^\infty E^2(r)\,r\,dr}\right]^{\frac{1}{2}}$$

These two parameters can be used for fibres with near-Gaussian fields. For such fibres, ω_j and ω_m can be used more accurately, particularly for predicting splice loss and bending sensitivity, respectively.

Direct measurement of MFD is made by the measurement of the transmitted near-field intensity distribution of a fibre 2—3 m long. The accuracy of this method depends on the lens perfection and dynamic range of the detector. An indirect measurement is the transverse offset method. If this fibre field is Gaussian,

$$P(d) = P_0 \exp |-d^2/2\omega_b{}^2|$$

obviously, the splice loss is determined at the same time. This method requires one fibre to be moved relative to the other at sub-micron accuracy laterally, while maintaining angular alignment. Several far-field measurement techniques can also be used. All involve measurement of the far field, and then either transform and curve-fit, or curve-fit directly. The technique is good for near-Gaussian field distribution. For non-Gaussian fields, the variable-aperture technique is used. This involves the use of a set of pre-determined apertures to intercept the far field. Data are taken of the power transmitted through these known apertures. The MFD is computed from the far-field pattern thus determined. Standardisation of MFD measurement has not yet been agreed.

7.11 References

1 KIM, E. M., and FRANZEN, D.: 'Measurement of far-field and near-field radiation patterns from optical fibers' NBS Tech. Note 1032 (US Government Printing Office, 1981)
2 WHITE, K. I.: 'The measurement of the refractive index profile of optical fibres by the refracted near-field technique', Fourth Eur. Conf. on Opt. Commun., Genoa, 1978, p. 146
3 REITZ, P.R.: 'Measuring optical waveguide attenuation, The LPS method', *Opt. Spectra*, 1981, p. 48.
4 COSTA, B., and SORDO, B.: 'Experimental study of optical fibres attenuation by modified backscattering technique, Third Eur. Conf. Opt. Commun., Munich, 1977
5 SAGE, A. P., and MELSA, J. L.: 'System identification' (Academic Press, New York, 1971), pp. 6–13
6 OKOSHI, T., and SASAKI, K., 'Precise measurement of the impulse response of an optical fibre', *Trans. IECE Jpn.*, 1978, **E61,** pp. 964–965
7 OKOSHI, T.: 'Optical fibers', (Academic Press, New York, 1982), pp. 243–245
8 Technical Staff of CSELT: 'Optical fiber communication' (McGraw Hill, New York, 1980), p. 205
9 AUFFRET, R., *et al.*: 'Vobulation technique applied to optical fibre transfer function measurement', 1st Eur. Conf. Opt. Fibre Commun., London, 1975, pp. 60–61
10 COHEN, L. G., and CHINLON LIN: 'Pulse delay measurements in the zero material dispersion wavelength region for optical fibres', *Appl. Opt.*, 1977, **16,** pp. 3136–3139
11 CONNELL, G. J., *et al.*: 'Measurement repeatability and comparison of real and equivalent step index (ESI) fibre profiles', Proc. Eur. Conf. Opt. Commun, Paper A IV 5, Cannes, 1982

Chapter 8

Optical fibre – Its prognosis and its economic impact

Optical fibre is an ideal transmission medium, and will remain a transmission medium par excellence. Its primary advantages of low loss, high bandwidth, small size, strength, flexibility and low cost, coupled with its special advantages of immunity to electromagnetic noise, safe and quiet operation, are not only unique but meet the exact requirements for transmission and other applications, such as for sensors and for signal-processing elements.

Originally envisaged for guiding coherent electromagnetic waves at optical wavelengths, fibre has taken over 20 years of development to reach the present state of readiness. At the same time, practical optical sources and detectors for the generation of coherent electromagnetic waves (stimulated emission of photons) and for their detection can be fabricated under controlled conditions. This has set the stage for optical fibre to meet and fulfil its full potential.

The prognosis for optical fibre is bright. As we progress into the post-industrial-revolution era we turn towards an information-intensive society where optical fibre is an essential component. Just as the motorways (autobahns or inter-state highways) permitted the development of the modern transportation system, optical fibre will become our information arterial highway, providing the transport system for our information traffic. Fibre extends its usefulness into the design of sensors, transducers and processors. Fibre elements have the obvious advantage of being compatible with the transmission medium. Thus, as fibre performance continues to improve and the infra-structure of related technology continues to build up, the range of applications of optical fibre will continue to widen.

Fibre loss of silica fibre is close to its assymptotic value of about 0·14 dB/km at 1·55 μm wavelength. The fine tuning of this loss is aimed at achieving a practical single-mode fibre with low dispersion over a spectral region and which can be manufactured at a low cost. More basic work on an exact electromagnetic solution for waveguides with frequency-dependent and profiled index distribution will improve fibre design. The basic work on glass-material-formation mechanisms and its relationship to scattering, radiation hardness, refractive index and nonlinear coefficients will substantially increase our ability to tailor the fibres to specific applications. Continuing work on new fibre-fabrication methods promises to reduce the basic cost further, so that single-mode fibre cost will one day be less than the cost of a pair of copper wires.

The sol gel process, in which the glass-constituent material starts in liquid form, and the mechanically shaped preform technique, which starts with soot material in powder form, are both volume-production techniques requiring low capital investment and a high-efficiency usage of raw materials. They hold new promise in reducing fibre cost. On the other hand, the improvement of traditional techniques also indicates an improvement by a factor of 2 to 4 times through up-scaling, and an increase in fabrication speed and yield.

Silica fibre, with its excellent mechanical characteristics together with the abundance of the raw material necessary for its production, will constitute the bulk of transmission fibres; these will be single-mode type. Except for repeaterless applications where repeater spacing must be as long as possible, i.e. for inter-continental trunking, trans-oceanic links or inter-island links having an intervening space of greater than 200 km, silica fibre is adequate. For trans-oceanic and island-hopping applications a low-loss fibre with a repeater-spacing bandwidth product of ~1000 GHz km is needed. This is indeed a challenge even if current materials work indicates that 10^{-3} dB/km is, in principle, possible for operation at 4—10 μm wavelength. Indeed, fluoride glasses, chacolgenide glasses and infra-red transmitting crystals should have the low losses as predicted if material structure can be controlled and impurities differentiated and removed. After that, the operating wavelength must be constrained to around the zero dispersion point with a nominally single-frequency source. The physical properties of the fibre must also be addressed. Most material is difficult to make, contains toxic components and is unstable, hygroscopic or susceptible to environmental attack.

Special single-mode fibres are needed for sensor applications where maintenance of polarisation is crucial. This requirement is also called for when fibres are used as signal-processing elements, although, when used as signal delay-lines, polarisation is not an issue. Multimode fibres are relegated to components roles. They may serve as short interconnection wires, and as fibre components where multimode facilitates the realisation of dispersive effects. This does not mean that the work on multimode fibre has been wasted. In fact, a thorough understanding of the characteristics of multimode system is needed for proper operation at single mode. Besides, the special properties of the multimode fibres can be used to make the fibre more versatile. For example, the large-diameter graded-index fibre is used as an excellent lens.

Other special single-mode fibres are needed for component applications. These may be fabricated with a variety of materials. A GeO_2 fibre can be used as a Raman amplifier. A low-temperature glass fibre can be the basis for making a range of microwave equivalent components such as couplers, attenuators, phase shifters etc. The material choice is to enable the fabrication of these components to be easier rather than ensuring, say, the low-loss characteristics. In component applications 3 dB/km translates to $<10^{-4}$ dB/10 cm, a negligible loss over such an unusually large distance encountered in a component.

For applications in a hostile environment and some sensor applications, silica fibre requires a coating of non-glassy material. SiN_2 and metals have been successfully used to coat the fibre to form a hermetic shield. However, depending on the coating method, the intrinsic strength of the fibre is lowered, presumably due to surface-crack formation. Coating uniformity can also cause excessive temperature-dependent bending losses. Research in this area includes controlling the deposition of material on the glass surface through controlling, either both or separately, the homogeneous and heterogeneous chemical reactions involved. This should improve deposition uniformity and adhesion, as well as broadening the range of materials usable for the fibre coating. Acoustically insensitive materials, materials with an opposite thermal expansion to glass, magnetic materials and electrostrictive material can allow different types of fibre sensors to be more readily realised and with increased sensitivity.

Fibre loss over the past, and projected to the future, is shown in Fig. 8.1. The application requirements for specific fibre characteristics are detailed in Fig. 8.2. The key milestones for fibre-performance projection expressed in bandwidth–repeater-spacing, together with the development of key components and techniques in associated technologies, is given in Fig. 8.3.

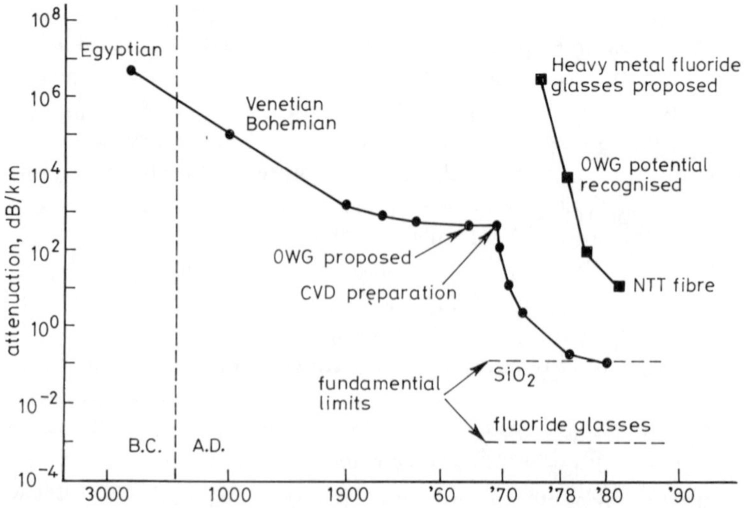

Fig. 8.1 *Fibre loss over the past and projected to the future*

characteristics / application	low loss	wide bandwidth	flexibility	strength/durability	small size	light weight	high temperature	radiation hardness	coherence
transmissions	•	•	•	•	•	•	•	•	•
sensors	•	•			•				•
signal processing	•	•			•				•

Fig. 8.2 *Application requirements for fibre characteristics*

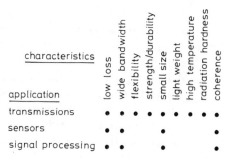

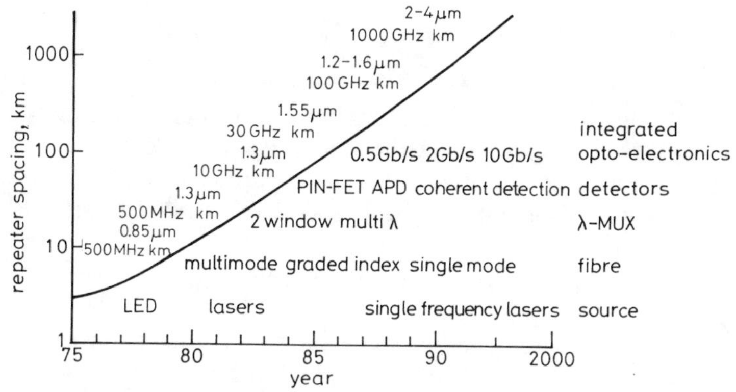

Fig. 8.3 *Key milestones for fibre-performance projection*

It is appropriate to note that the bandwidth–distance product is capable of handling 1000 GHz over 1 km. This raises the issue that the signal-processing speed at the terminals should be very substantially increased in order to fully utilise, if needed, the bandwidth provided by the fibre. Currently fibre systems have been designed to handle signals up to 2·24 Gbs. Highest bit-rate signal transmission demonstrated is below 10 Gb/s. Two pertinent questions are: (i) Is 10^{12} b/s or terabit rate possible for future optoelectronic components; (ii) If 10^{12} b/s can indeed be attained, how should the design of signal-handling systems be changed to take the advantage of such high signal-processing speed. These questions are particularly timely since the transmission capabilities of the fibre are adequate to cater for these signalling rates. Moreover, several applications indicate that the demand on information capacity, transmission and processing rate can make use of such high signalling rates. For example, if video phones at 70 Mb/s per channel are as densely installed as telephones at 64 kb/s per channel today, the transmission requirement would be 1000 times the current trunk-systems transmission rates of 560 Mb/s; hence, 560 Gb/s are needed. Certainly one should also revise the strategy of all signal-processing methods, which hitherto have been designed to conserve bandwidth. With bandwidth available in abundance, signal-processing schemes which are wasteful in bandwidth, but which simplify the overall system design and are more cost effective, should be entertained.

Not surprisingly, optical fibre in an information society is stimulating research into high-speed optoelectronic components. It is speculated that the combination of electrons and photons used appropriately can result in effective signal-processing rates beyond what can be achieved in pure electronics for both silicon and GaAs technologies (see Fig. 8.4).

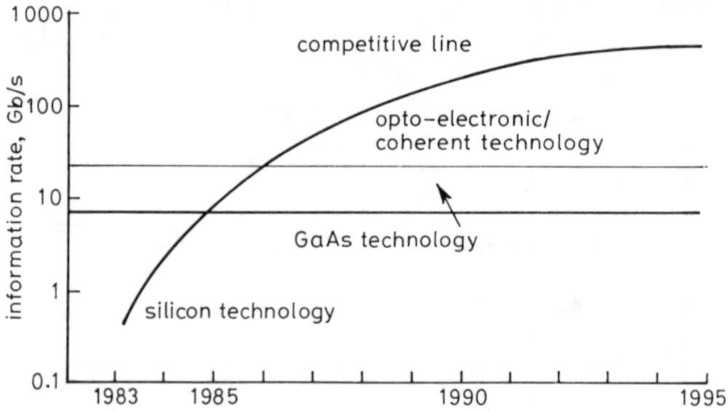

Fig. 8.4 *Projection of high bit-rate technologies*

The evolution of optoelectronics is taking place. Advances in electronic-material fabrication techniques, such as MBE and MOCVD, are making new semiconductors with 'tailored' properties possible. This is sometimes referred to as bandgap engineering. Study of the electron-transport mechanism in semiconductor material and the availability of ultra-fast optical-spectroscopy diagnostic techniques are together causing significant steps to be made toward realising truly integrated optoelectronic functions, which will evolve into system building blocks for ultra-fast systems. In the meantime broadband communication in integrated or separate networks is being promoted, particularly strongly in Europe. Fig. 8.5 raises the scenario of B-ISDN, or separate networks. Fig. 8.6 shows trial systems being installed in Europe.

Field trials are providing opportunities to test not only technical feasibility, but also customer reactions to various experimental services. These activities will increase significantly in the near future when the benefits of such a system begin to be perceived, not in terms of the extension of conventional communication services such as telephones to video phones, data to ISDN etc., but in terms of new services based on fulfilling real information needs. Already stock-market information, instant banking transactions and legal

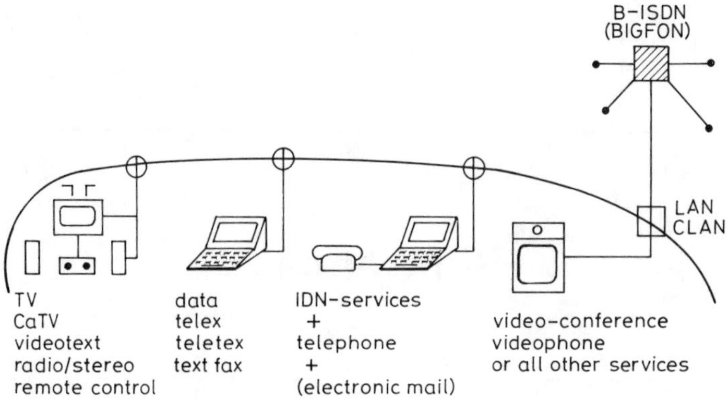

Fig 8.5 *B-ISDN or separate networks*

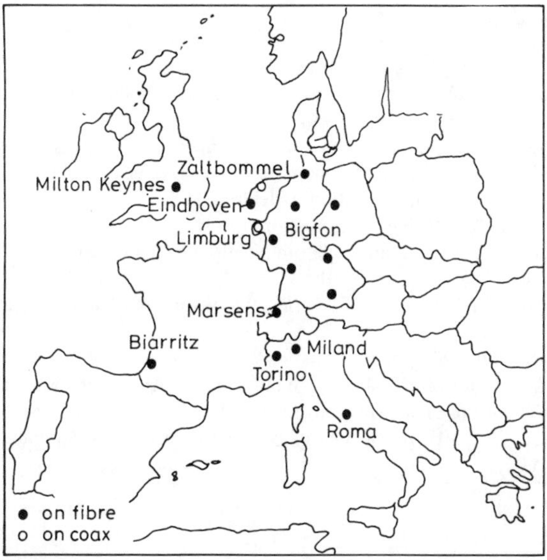

Fig. 8.6 *Map of European experimental broadband integrated service network*

search services have created a strong demand. New services backed by good interactive data bases can provide the competitive edge in doing business in all sectors. We must move closer to being able to master our data and knowledge resources.

Chapter 9

References

9.1 Books

MIDWINTER, J. E.: 'Optical fibers for transmission' (John Wiley, 1979)
WOLF, H. F. (Ed.): 'Handbook of fiber optics theory and applications' (Garland STPM Press, 1979)
KAO, C. K.: 'Optical fiber systems, technology design and applications' (McGraw Hill, 1982)
SANDBANK, C. P. (Ed.): 'Optical fibre communication systems' (John Wiley, 1980)
KEISER, G.: 'Optical fiber communication' (McGraw Hill, 1983)
BASCH, E. E. (Ed.): 'Optical fiber transmission' (Howard W. Sams, 1987)
Technical Staff of CSELT 'Optical fiber communications' (McGraw Hill, 1987)
OKOSHI, T.: 'Optical fibers' (Academic Press, 1982)
MILLER, S. E., and CHYNOWETH, A. G. (Ed.): 'Optical fiber telecommunication' (Academic Press, 1979)
PERSONICK, S. D.: 'Optical fiber transmission systems' (Plenum Press, 1981)
MARCUSE, D.: 'Principles of optical fiber management' (Academic Press, 1987)
MARCUSE, D.: 'Theory of dielectric optical waveguides' (Academic Press, 1974)
SNYDER, A. W., and LOVE, J. D.: 'Optical waveguide theory' (Chapman & Hall, 1983)
YARIV, A., and YEH, P.: 'Optical waves in crystals' (John Wiley, 1984)

9.2 Miscellaneous

GOODMAN, C. H. L.: 'Devices and materials for 4 μm-band fiber optical communication', *J. Solid State & Electron Devices*, 1978, **2**, pp. 129–137
SCHULTZ, P. C.: 'Progress in optical waveguide process and materials', *Appl. Opt.*, 1979, **18**, pp. 3684–3693
GIALLORENZI, T. G.: 'Optical communications research and technology: Fiber optics', *Proc. IEEE*, 1978, **66**, pp. 744–780
MURATA, H.: 'Low-loss single-mode fiber development and splicing research in Japan', *IEEE J. Quantum Electron.*, 1981, **QE-17**, pp. 835–849
GLOGE, D.: 'Optical fiber packaging and its influence on fiber straightness and loss', *Bell Syst. Tech. J.*, Feb. 1975, pp. 245–262
INAGAKI, N., EDAHIRO, T., and NAKAHARA, M.: 'Recent progress in VAD fiber fabrication process', *Japan J. Appl. Phys.*, 1981, **20**, Suppl. 20-1, pp. 175–180

INAGAKI, N.: 'Recent progress in glass fibers for optical communication', *Japan J. Appl. Phys.*, 1981, **20,** pp. 1347–1360

IZAWA, T., and INAGAKI, N.: 'Materials and processes for fiber preform fabrication – Vapor phase axial deposition', *Proc. IEEE*, 1980, **68,** pp. 1184–1187

FRIEBELE, E. J., SIGEL, G. H., Jr., and GINGERICH, M. E.: 'Radiation response of fiber optic waveguides in the 0·4 to 1·7 μm region', *IEEE Trans.*, 1978, **NS-25,** pp. 1261–1266

KAO, K. C., and HOCKHAM, G. A.: 'Dielectric fiber surface waveguides for optical frequencies', *Proc. IEE*, 1966, **117,** pp. 1151–1158

WALKER, K. L., GEYLING, F. T., and NAGEL, S. R.: 'Thermophoretic deposition of small particles in the modified chemical vapor deposition (MCVD) process', *J. Am. Ceramic Soc.*, 1980, **63,** pp. 552–558

SMITH, R. G.: 'Optical power handling capacity of low loss optical fibers as determined by stimulated Raman and Brillouin scattering', *Appl. Opt.*, 1972, **11,** pp. 2489–2494

Index